Animal Body Fluids and their Regulation

Animal Body Fluids and their Regulation

by

A. P. M. LOCKWOOD M.A., PH.D.
Lecturer in Zoology, University of Southampton
Formerly Research Fellow, Trinity College, Cambridge

HARVARD UNIVERSITY PRESS
CAMBRIDGE, MASSACHUSETTS
1964

Printed in Great Britain

Preface

IN recent years there has been a considerable volume of research directed at attempts to relate the physiology of animals to the special conditions found in their particular environmental niches. In consequence there has been an increasing emphasis placed on a physiological approach to biological problems in teaching at both the school and undergraduate level. Some aspects of physiology, as for example respiration and nervous conduction, have been summarised in a manner suitable for students in these categories. Others, including the mechanisms by which animals regulate their salt and water balance, are not comprehensively covered at this level. This is unfortunate not only because the maintenance of the correct amount of water and minerals in the body is of fundamental importance in the operation of all other metabolic processes but also because some complex physiological mechanisms are apparently based on processes originally used for regulating cellular water and ion content but now modified for other purposes. For example the ability of nerves to conduct impulses is derived from the capacity of cells to extrude sodium and retain potassium.

It is hoped that this little book will provide a general account of the mechanisms involved in the regulation of the water and ion content of blood and cells and will show some of the modifications in the general physiology necessary if life is to be sustained in a variety of different habitats.

My sincere thanks are due to my colleagues, Drs J. W. L. Beament, G. M. Hughes, G. Shelton, R. V. Short and J. E. Treherne, who have read certain chapters and helped me to avoid a number of errors. They are in no way to blame for any that may still remain.

June 1963 A.P.M.L.

Acknowledgements

I am most grateful to all the people who have assisted me directly or indirectly in the production of this book. Deserving of special mention are the editor of *The Scholarship Series*, Mr W. H. Dowdeswell, for his advice on the original typescript, my colleagues, Drs J. W. L. Beament, G. M. Hughes, G. Shelton, R. V. Short and J. E. Treherne, who have helped me to avoid a number of errors, and Mr H. MacGibbon of Heinemann Educational Books Ltd for all his helpful guidance.

I would also like to thank the many authors whose work has contributed to this volume and in particular Profs O. Kinne and K. Schmidt-Nielsen and Drs R. M. Chew, M. S. Gordon, W. J. Gross, J. A. Ramsay, J. D. Robertson and J. Shaw for permission to make use of diagrams or tables from their publications.

I would also like to acknowledge that the following figures are based, in a more or less modified form, on diagrams first appearing in the journals listed below, figs. 2, 11, 18 and part of 29 from the Journal of Experimental Biology; fig. 13 from the American Journal of Physiology; fig. 12 from the Journal of Cellular and Comparative Physiology; fig. 28 from Biological Reviews and parts of figs. 3 and 14 from the Biological Bulletin.

A. P. M. L.

Contents

1

Introduction

Animal life is believed to have originated in the sea, but species have evolved which have subsequently established themselves in fresh water and on land. 'Invasions' of fresh water have occurred many times but nevertheless, though this medium carries a high density of individuals, it compares unfavourably with the sea in the number of species present. This relative paucity of species gives an indication that certain difficulties are involved in the colonisation of fresh water since every species of animal will tend to extend its range to the limits imposed by its physiological adaptability and by competition with other animals. To understand why there should be this difficulty we must consider the various major differences between the sea and fresh waters.

Both the physical and chemical properties of sea water are relatively stable in any one place. Factors such as the concentration of minerals, temperature, density, acidity and

TABLE 1. Ion concentrations in parts per million

	Na	*K*	*Ca*	*Mg*	*Cl*	SO_4
Sea Water	10,700	380	420	1,320	19,320	2,700
London Tap Water	20	5	97	4	27	49
Ennerdale Lake Water	5·8	–	0·8	0·7	5·7	4·6

temperature vary rather little during the year and such changes as there are tend to be slow. Hence there is adequate time for marine animals to make any necessary physiological adjustments. Conditions in fresh waters are much less constant. The total mineral concentration is only about 1/1000th to 1/100th that of sea water and is very variable both in concentration and ionic composition. Annual temperature fluctuations are often large and in small bodies of fresh water even the daily change may be considerable. In well mixed water the oxygen content may exceed that of the sea, but in swamps and stagnant pools very low levels may be present. The pH is very variable.

Marine animals are ill-fitted for life in such an exacting medium and before a marine form can extend its range even into brackish water evolution must first bring about modifications in its biochemistry, physiology, behaviour, and micro-anatomy. Indeed, the nature of the changes involved in becoming fully adapted to life in fresh water is usually sufficiently drastic to preclude any possibility of return to the sea. Animals such as the eel and salmon which migrate freely between rivers and the sea are rare exceptions to this general rule and possess somewhat unusual attributes. However, though there are few species able to move from one medium to the other, there are indications that many of the animal groups now living in the sea or fresh water were derived from forms which originally lived in the other medium. During the evolution of the vertebrates there have apparently been many such migrations from the sea to fresh water and vice versa.

It is probable that the primitive chordates lived in the sea but that early in the history of the vertebrates some of the jaw-less fishes (Agnatha) successfully developed the capacity to live in fresh water. The jawed fish are presumed to have arisen from these agnathans in fresh water but the forms giving rise to the elasmobranchs then remigrated back to the sea and were followed at a later stage by the ancestors of the modern marine teleosts. As a further complication there are indications that a number of fresh water teleosts today, including some pipe fish, are derived from the marine teleosts and hence represent a further remigration back into fresh water.

The higher vertebrate groups arose from non-marine forms but, with the exception of the amphibia, all now include some species which live in or close to the sea. Familar examples include the marine turtles, sea-snakes, penguins, gulls, and mammals such as the whales, seals and dolphins. (Fig. 1.)

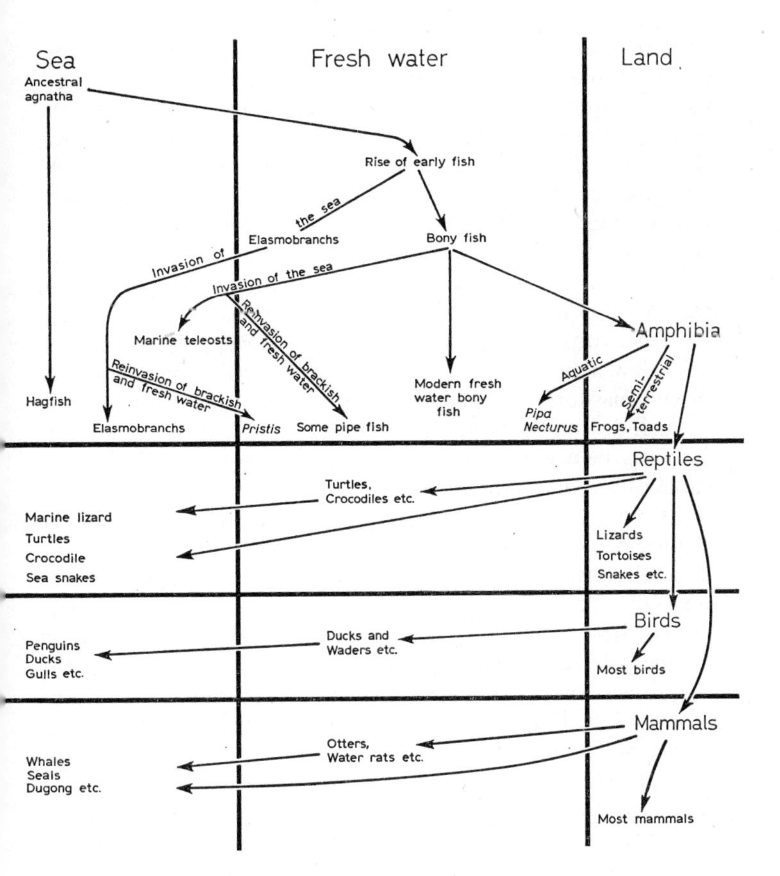

FIG. 1. An outline of some of the probable movements between different environments which have occurred during the evolution of the vertebrates. (In the case of the sections on reptiles, birds and mammals no phylogenetic significance is intended, the arrows indicating only the general derivation of fresh water and marine forms from terrestrial ancestors.)

Similar movements between the sea, fresh water and land have also occurred in the invertebrate phyla. As in the chordates it is probable that the majority of forms now living on land were derived ultimately from fresh water ancestors. Invasion of the land directly across the sea shore is thought to have been rare, though it is likely that this is the route followed by the land crabs and possibly also by the isopods (woodlice) and some molluscs. Fresh water has been colonised many times by invertebrates from the sea but has also been recolonised by species belonging to essentially terrestrial groups. The pulmonate molluscs and the aquatic insects are examples of the latter movements. Evidence for recolonisation of the sea is less common but there are a few marine insects and a number of pulmonate molluscs on the sea shore both of which must have non-marine ancestors.

Such evolutionary moves from one environment to another can only be accomplished when the animals involved have the capacity to maintain their body fluid composition and concentration in the new medium at a level compatible with the requirements of their cells. Animal cells have thin and permeable cell walls and water is gained or lost if the medium bathing them is less or more concentrated than the cell contents. This fact may readily be verified by placing a drop of vertebrate blood in distilled water or 5 per cent salt solution. In distilled water the corpuscles swell and release haemoglobin and in the concentrated medium they shrink and the cell membrane becomes wrinkled. Such gross changes in cell volume are incompatible with normal physiological function and must be avoided in an intact animal. It is important therefore that cells are (1) bathed by a medium with the same osmotic concentration as themselves, (2) able to excrete water as rapidly as it enters or (3) able to adjust their own internal concentration to match any variation in that of the blood.

The nature of the problems involved in providing cells with a suitable bathing medium depends on the conditions imposed by the environment. The concentration and composition of sea water is close to the requirements of the cells of most marine invertebrates and these animals have blood which is essentially similar to sea water in composition. A medium with an ion concentration as low as that of fresh water is too dilute for the functioning of the protoplasm of higher organisms and hence animals living in this medium must maintain their body fluids at a concentration above that of the environment. However, when an animal's body fluids are more concentrated than the environment it tends to take up water by osmosis and to lose ions by diffusion down the concentration gradient. In theory the body surface might be rendered sufficiently impermeable to prevent such movements of water and ions, but in practice only a few air-breathing aquatic forms are able to approximate to this ideal solution. Animals with an aquatic mode of respiration could not decrease the permeability of their whole body surface without at the same time restricting the rate of diffusion of oxygen. Such forms must, therefore, have the means both to eliminate water taken up by osmosis and to replace lost ions. The water is removed as urine and inorganic ions are actively transported from the medium to the blood.

The rate at which water is taken up and ions lost varies with the concentration gradient and the permeability of the body surface. The capacity merely to take up ions at a fixed rate would therefore be inadequate, there must also be a means of regulating the rate of transport. This regulation involves the intervention of sensory, feedback, and control mechanisms.

In some environments the blood concentration fluctuates when the medium changes despite variations in the rate of active transport. Brackish-water animals in particular show wide variation in blood concentration. Any such change must be matched by modification of the solute content of the body cells if they are not to take up or lose water and vary in volume.

The above outline indicates that a whole gamut of physiological conditions must be satisfied before the colonisation of non-marine aquatic habitats is possible.

However, even if the adults can tolerate dilute media, a species cannot become permanently established in fresh water unless the young stages are also viable. Many marine species produce small, unprotected eggs which are broadcast into the sea. The young stages then move up and maintain themselves in the rich planktonic feeding zone near the surface. Such broadcast release of young in fresh water would be disastrous. Not only would the young have to expend considerably more energy in remaining near the surface in the less dense medium but they would also be liable to be swept downstream wherever there were currents. The young of fresh water animals must therefore be protected and provided with food until large enough to fend for themselves. To this end the eggs of most fresh water forms are usually larger than those of their marine relatives and the young tend to hatch at an advanced state of development.

Animals must also be preadapted to life on land. Before leaving the water they must be able to breath atmospheric oxygen and, what concerns us most here, maintain the blood concentration by restricting loss of water in the urine and by evaporation.

New behaviour patterns must be developed to enable terrestrial forms to seek out and ingest the salts and water they require and to avoid areas of low humidity where evaporation would be rapid. Modification of the metabolism of nitrogen is also necessary to ensure toxic waste products are eliminated in the little water available for the production of urine.

The following pages are devoted to a consideration of the factors of various environments which influence the concentration and composition of animal body fluids and of the various means adopted by different organisms to regulate their internal composition.

Osmosis and other terms

Hydrostatic pressure and osmosis are the most important factors in the movement of water between different compartments in the body, and between the body and aquatic surroundings.

The molecules of an aqueous solution are in constant thermal motion. When substances are dissolved in water the activity of the water molecules is reduced as a result of their interaction with those of the solute. If water containing a solute is separated from pure water by a membrane permeable to water but not to the solution molecules (semipermeable) the water molecules have an average activity higher on one side of the membrane than on the other. Molecules from the pure water traverse the membrane with greater frequency than those from the solution. The movements of water in the two directions across the membrane are termed fluxes and, since there is a quantitative difference between these fluxes there is a net movement of water molecules from the pure water to the solution. This passage of water through a semipermeable membrane from a lower to a higher concentration of a solute is termed osmosis.

The movement of water in this manner will establish a difference in surface level and hence of hydrostatic pressure between the two solutions. This pressure tends to oppose the entry of water and, when of sufficient magnitude, prevents further net movement of water across the membrane. The hydrostatic pressure required to prevent the net entry of solvent across a semipermeable membrane is termed the osmotic pressure of the solution. It is important to note that a solution does not exert a pressure on the side of a solid container by virtue of its osmotic pressure. The pressure exerted on the walls of a beaker by 100 ccs of distilled water and 100 ccs of a saturated salt solution will, for all practical purposes, be the same.

The total concentration of a solution is usually expressed as a measure of the number of particles of solute in 1 litre or kilogram of water. Concentration is measured in mols/litre. A molar solution is one in which the molecular weight of a substance in grams (gram mol) is dissolved and made up to 1 litre with water. A gram mol dissolved in 1 kilogram of water is called a molal solution. In dilute solutions of substances with a low molecular weight there is little difference between molal and molar solutions; but where substances of high molecular weights are present in solution, as in most body fluids, the difference may be considerable.

The laws obeyed by solutions are similar to those of gases. 1 gram mol of gas fills a volume of 22·4 litres at 0° C. and 1 atmosphere and 1 gram mol of a non-electrolyte (a substance that does not break down into ionised particles when it goes into solution) dissolved in 1 kilogram of water at 0° C. has an osmotic pressure of 22·4 atmospheres. Since the osmotic pressure of a solution depends on the relation between the number of solute and water molecules the number of particles produced when a substance is dissolved will govern the osmotic pressure. A solution of 0·01 gram mol of sodium chloride in a kilogram of water has almost twice the osmotic pressure of a solution of 0·01 gram mol of glucose similarly dissolved since on solution it ionises into two particles, Na^+ and Cl^-. Similarly, substances such as $MgCl_2$ which ionise into three particles have an even higher osmotic pressure per gram mol. The osmotic pressure of electrolytes is not linearly related to concentration. At high concentrations the osmotic pressure is lower than might be expected, probably because of mutual interaction between positive and negative particles.

Solutions of similar osmotic pressure have certain linked properties in common. Their vapour pressures, freezing points and boiling points are all the same. Solutions which have such similar properties are said to be *isosmotic*. The term *isotonic* has been loosely used in the same context but should be retained for cases where it has been shown that there is no net movement of water across a membrane separating two solutions. Biological membranes are rarely semipermeable and two isosmotic solutions separated by such membranes are not always isotonic.

A solution which is less concentrated than another is termed *hyposmotic* and one which is more concentrated is *hyperosmotic*. The corresponding terms for tonicity are *hypotonic* and *hypertonic*.

Osmotic pressure measurements are expressed in a variety of ways: as mols/l. of a non-electrolyte (osmoles), as mols/1 of a solution of NaCl, in atmospheres, or in terms of the depression of freezing point (△° C.). E.g. the concentration of a 1% solution of NaCl may be expressed as 171 mM/1. NaCl, 0·32 osmoles, △ 0·595° C. or 7·17 atmospheres.

The concentration of ions is usually given in terms of the equivalent weight in 1 litre (Equivs. /1). As most biological materials are fairly dilute the units most commonly used are milli mols per litre (mM/1), milli equivs (mE/1) etc, which are 1000th of a mol or equivalent per litre.

Animals which tolerate a wide range of salinities are termed *euryhaline*, those that are restricted to a narrow salinity band, *stenohaline*.

2

Invertebrate Body Fluid Regulation

The same basic physiological processes seem to be used in the regulation of the body fluid concentration and composition in all animals but the extent to which special reliance is placed on particular mechanisms depends both on the group of animals and on the nature of the environment.

AQUATIC INVERTEBRATES

Marine species

Most marine invertebrates have body surfaces which are rather permeable to ions and water, and their body fluids are isotonic with sea water. If they are transferred to diluted sea water ions are lost by diffusion and water is taken up by osmosis. The passage of water usually occurs more rapidly than the loss of ions so soft bodied forms such as annelids and molluscs swell initially when placed in dilute media. Subsequently the weight returns towards the normal level (Fig. 2). Weight changes are less extensive in crustacea as the cuticle prevents gross expansion, but the end result is the same as in the annelids and molluscs—a marked decrease in blood concentration. Such forms can survive variations in their medium only to the extent that their cells will tolerate dilution. In general such tolerance at the cellular level is very limited in offshore marine animals. Thus by experiment it has been found that the crabs *Maia* and *Hyas* will only survive in media more concentrated than about 75–80 per cent sea water. Even so such experiments tend to give an exaggerated conception of the normal limits of tolerance of such forms since the effects of salinity on the competitive powers of the animal are ruled out. In natural conditions quite small variations in salinity may grossly affect the fauna. Thus the waters of the Skagerrak contain only about 25 per cent of the species present in the North Sea even though the difference in salinity between the media is only 10 per cent sea water.

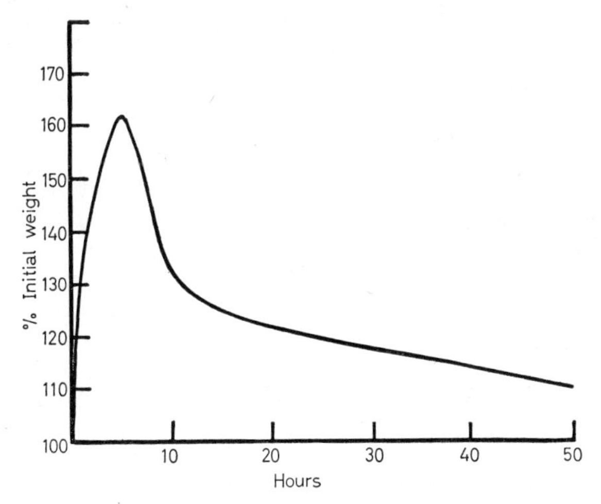

FIG. 2. Weight changes in the polychaete, *Nereis diversicolor*, on transference from 100 to 20 per cent sea water. To show that a soft-bodied, permeable animal at first takes up water when transferred to a dilute medium but later returns to near its original weight. (After Ellis)

Brackish water species

Isosmotic species. Almost all the animals which live in brackish waters today are derived from marine forms. They are physiologically advanced, however, in that either the general body cells withstand changes in blood concentration or the blood can be maintained hyperosmotic to the medium. Few species depend entirely on the salinity tolerance of their cells, and the number of animals which are isosmotic with their medium over a wide range of concentration is consequently limited. However, the common mussel, *Mytilus*, and the lugworm, *Arenicola*, penetrate into brackish water and, in the Baltic, may be found in concentrations as low as 15 per cent and 30 per cent sea water respectively. Both forms have blood which is in fact effectively isotonic with the medium and hence their osmotic regulation is essentially confined to the level of the general body cells. Forms living in brackish water which have less permeable body surfaces can maintain the blood hypertonic to the medium over a considerable range of concentration and thus limit the extent of regulation required at the cellular level.

Hyperosmotic forms. Most arthropods which occur naturally in brackish water maintain their blood hypertonic to the medium and hence cannot avoid loss of ions by diffusion and in the urine. These losses are made good by an active uptake of ions from the medium but as such uptake involves the expenditure of energy the rate of uptake has to be kept as low as possible by minimising the rate of loss. Consequently forms which are hypertonic to the medium tend to decrease the work they have to do by (1) making the body surface less permeable, and (2) maintaining a small rather than a large concentration gradient between blood and medium.

Crustacean species show a good correlation between the permeability of their cuticle and the environment in which they live. Isosmotic forms from the sea have more permeable surfaces than estuarine species and these in turn are more permeable than fresh water species. Similarly littoral species from the upper part of the tidal range, which are more likely to be exposed to loss of water by evaporation than forms from the low and sublittoral regions, have lower permeabilities than the latter (Fig. 3). The necessity of allowing an adequate diffusion of oxygen across the body surface limits the extent to which brackish and fresh water forms can decrease their permeability, and hence these forms cannot make themselves so impermeable as to be able to maintain a constant blood concentration. Hence despite their restricted permeability brackish water species vary the blood concentration as that of their medium changes. The blood

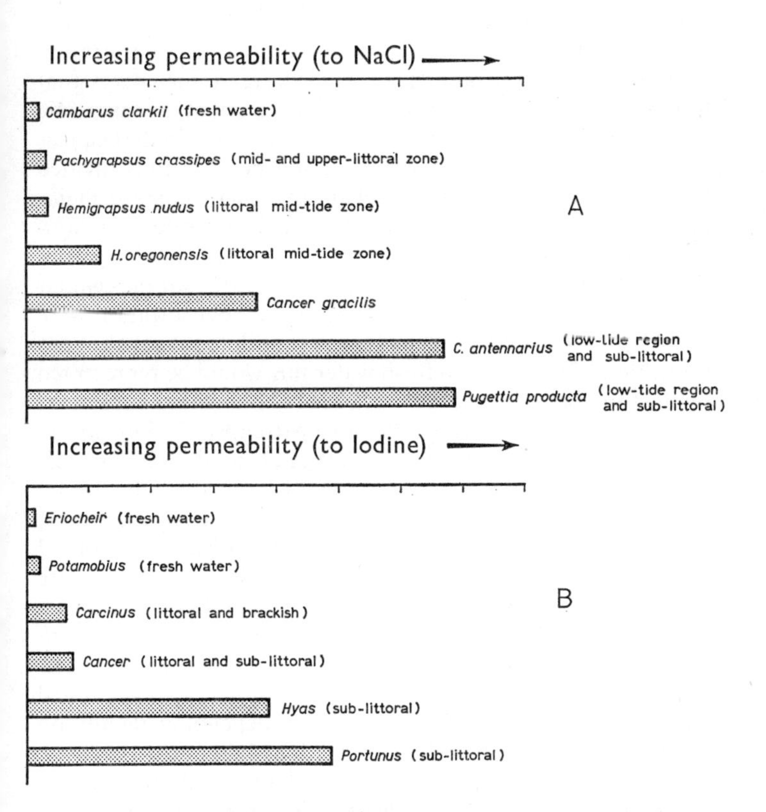

FIG. 3. (a) Differences in the salt permeability of the cuticle of crustacea from different environments. Note that the forms exposed to osmoregulatory stress have the least permeable cuticles. (After Gross)
(b) Differences in the permeability of the cuticle of marine, brackish and fresh-water crustacea to iodine. The forms exposed to osmotic stress have lower permeabilities than the marine forms. (After Nagel)

concentration changes in the same sense as that of the medium though not to the same degree. The fall in blood concentration when the medium is diluted serves to decrease the energy that has to be expended on the transport of ions, but at the same time it means that the body cells must be able to adjust to the changes involved. The overall regulation of the body is thus divided between body cells and the mechanisms responsible for maintaining the stability of the blood. Fig. 4 illustrates diagrammatically the effect on the concentration of the blood of variations in the division of labour between regulation at the cellular level and at the body surface. If the blood concentration were to be kept constant as the medium were diluted from the concentration of sea water to fresh water this would be represented by the top line and no adjustment would be required at the level of the general body cells. At the other extreme if the blood were isosmotic with the medium over a range of salinities it would lie along the diagonal line and the animal's survival would depend on the capacity of its cells to tolerate dilution. In practice most animals that live in brackish water compromise and have blood concentrations intermediate between these limits. The cells are thus protected from extreme changes without the animals having to expend an excessive amount of energy in transporting ions to maintain a high gradient between blood and medium. The cross-hatched area in Fig. 4 indicates the approximate limits of blood concentration that are found. Animals depending mainly on cell regulation have blood concentrations lying near the lower border of this area, e.g. *Arenicola* and *Mytilus*. Animals relying primarily on active transport have values near the upper edge, e.g. *Carcinus*, the shore crab. It should be noted, however, that few species and none of the above, can survive over the whole range from fresh water to sea water.

In some cases the emphasis placed on cell tolerance or active transport depends on the concentration of the medium. Thus the amphipod, *Gammarus duebeni*, relies on active transport when the medium is dilute and on cell regulation when it is concentrated. In the range fresh water to 50 per cent sea water the blood composition varies rather little but in the range 50 per cent to 200 per cent sea water the blood is almost isosmotic with the medium.

Gammarus duebeni lives primarily in supra-littoral pools and estuarine backwaters both of which are liable to wide fluctuations as a result of evaporation or inundation by fresh water.

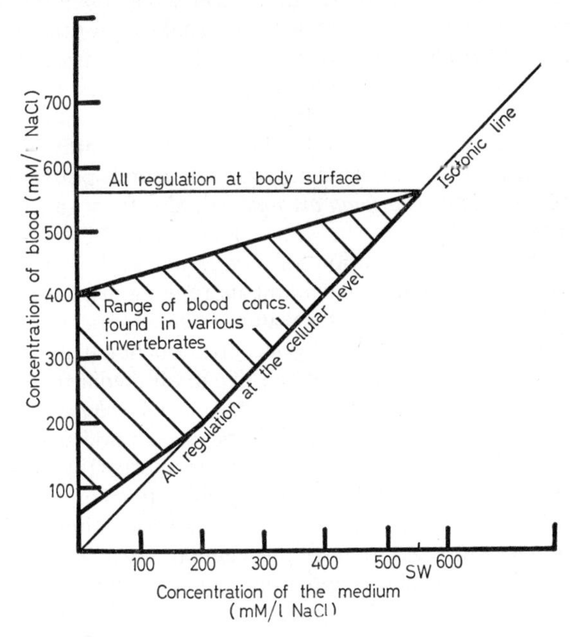

FIG. 4. The relation between osmotic regulation at the body surface and at the level of the general cells of the body in brackish water animals maintaining various gradients of concentration between their blood and the medium.

Its wide powers of salinity tolerance are therefore necessary. Yet another form of regulation of the blood is found in other animals which live in similarly rigorous environments. This involves the maintenance of the blood hyperosmotic to the medium when the concentration of the latter is low and hyposmotic to the medium when the concentration is high.

Hyper- and Hyposmotic regulation. When the blood is hyperosmotic to the medium ions must be taken up from the medium and water extruded. When the blood is hyposmotic to the medium regulation necessitates the uptake of water and the extrusion of excess ions. Water is probably obtained by such forms through drinking the medium. Examples of animals showing this form of regulation include the brine shrimp, *Artemia salina*, from inland saline lakes, the prawn, *Palaemonetes varians*, from estuaries and salt marsh pools and the isopod, *Gnorimosphaeroma oregonensis*, also from salt marshes and estuaries.

The blood concentration of *Palaemonetes* is held almost constant when the animal is in media ranging in concentration from about 2–110 per cent sea water. *Artemia* does not maintain quite such a constant blood concentration but it has a much wider range of tolerance as it can live in all media from 10 per cent sea water to crystallising brine. *Gnorimosphaeroma* maintains a comparatively small gradient between blood and medium both when hyperosmotic and when hyposmotic (Fig. 5). Hence these forms which can be both hyper- and hyposmotic, like those that are only hyperosmotic, also show variations in their dependence on regulation at the cellular level. *Palaemonetes* requires little cellular regulation, *Gnorimosphaeroma* rather more.

Fresh water species

The extension of the regulatory abilities of brackish water animals have enabled the colonisation of fresh water.

Many of the species of large body size such as the crabs, *Potamon niloticus* and *Eriocheir sinensis*, maintain in fresh water a blood concentration which is almost as high as that of crabs such as *Carcinus* which are confined to brackish water. Further, and again like their brackish water relatives, they form urine isosmotic with the blood. It might be expected therefore that they would have to perform much work to maintain the blood equilibrium. They avoid this necessity by being extremely impermeable, their water turnover representing only a small pro-

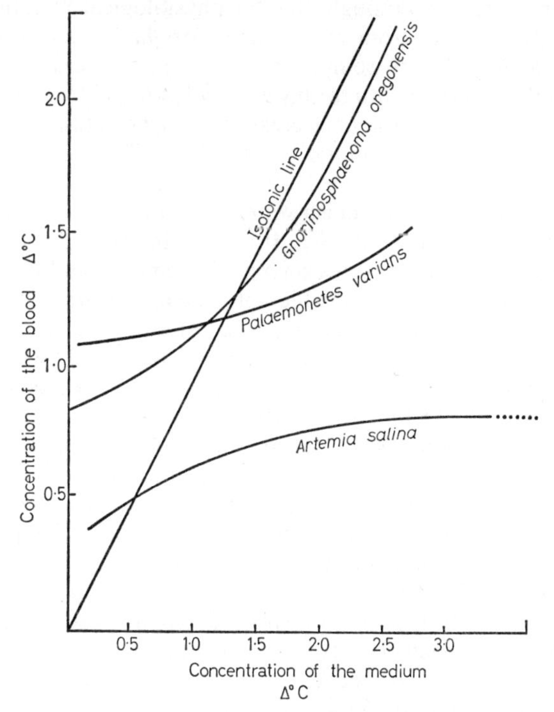

FIG. 5. The relationship between blood concentration and the concentration of the medium in three animals whose body fluids are maintained hypotonic to the medium when the latter is highly concentrated. To show the variation between species in the concentration gradients maintained.

portion of the body weight per day. More permeable or smaller forms in fresh water cannot afford to maintain such a large concentration gradient as the crabs and have lower blood concentrations. In addition, all the more permeable forms are found to restrict ion loss during the removal of excess water by producing urine hyposmotic to the blood.

At one time it was thought that the physiological evolution of fresh water species from marine ancestors had taken place in three stages: (1) decrease in permeability and increased rate of ion uptake; (2) entry into fresh water with a high blood concentration followed by a later decrease in concentration; (3) production of urine hyposmotic to the blood. The first of these stages is certainly correct for crustacea as we have already noted that species found in brackish water have a lower permeability than forms confined to the sea. It is most improbable however that highly permeable forms such as the annelids and molluscs could ever have maintained a high blood concentration in dilute media, and in these groups tolerance at the cellular level of decrease in the blood concentration and increased powers of active uptake of ions must have been contemporaneous.* A former hypothesis that the capacity to produce urine hyposmotic to the blood was developed only after the colonisation of fresh water must also be discarded. The brackish water amphipod *Gammarus duebeni* produces dilute urine when it is in media with a salinity less than that of 50 per cent sea water. The crabs, *Eriocheir*, and *Potamon* lose only about 10 per cent and 1 per cent respectively of their daily salt loss in the urine and hence there would be little point in their producing hyposmotic urine. In *Gammarus duebeni* however when the urine is isosmotic with the blood some 80 per cent of the salt loss is via this route. Production of urine hyposmotic to the blood when the animal is in very dilute media thus effects a very considerable conservation of ions that would otherwise be lost. When *G. duebeni* is in fresh water the daily urine production is estimated to be over 70 per cent of the body weight, a very high rate in comparison with the fresh water crabs. However, it must be remembered that a copious production of urine by small forms does not

* The process of complete adaptation of crustacea to fresh water probably does entail some lowering of the blood concentration after their initial penetration of this medium from brackish water.

necessarily imply that they are any more permeable to water per unit surface area than larger forms such as the crabs; the difference may be due to the effect of size. Fully grown crabs weigh 100 grams or more, some 1,000 times the weight of a *Gammarus*. Since the ratio of surface to volume increases with decreasing size, a small animal will tend to have a greater water

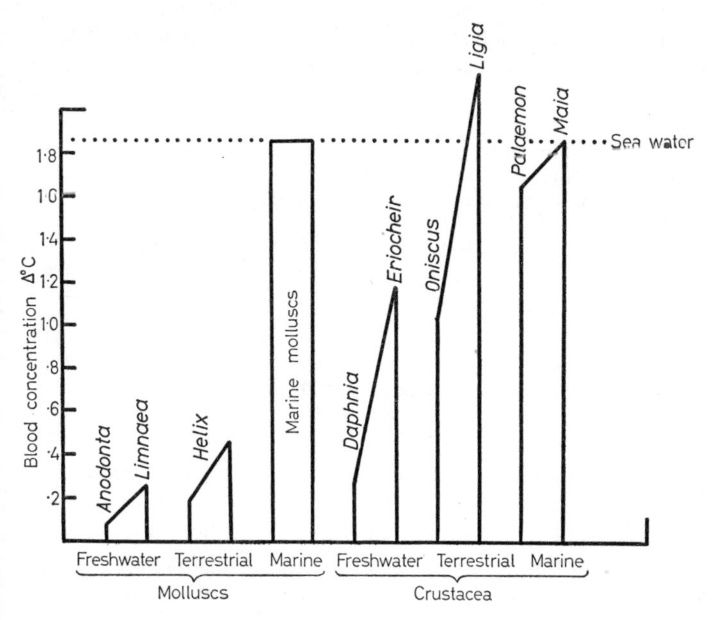

FIG. 6. The concentration of the blood of molluscs and crustacea from various environments. Note that the more permeable forms, the molluscs, have lower blood concentrations both in fresh water and on land than the less permeable crustacea.

turnover for a given surface permeability and concentration gradient than a large form (provided of course that the permeable part of the surface is not corrugated so as to keep the surface to volume ratio constant). The ion turnover of small forms will also be larger than in species of bigger body size and hence the amount of work which must be done per unit mass of body tissue to maintain a given concentration between blood and medium will increase with decreasing size. Thus small animals, as well as those which have very permeable surfaces, will gain an advantage if they decrease the blood concentration and produce hyposmotic urine when the medium is very dilute. It is notable that all the smaller species of crustacea living in fresh water have lower blood concentrations than forms of larger body size (Table 2), (Fig. 6).

TABLE 2. The effect of body size on water turnover in fresh water crustacea. Note that the faster the water turnover and the smaller the adult size the lower is the normal blood concentration.

Species	*Approx. wt.*	*Blood Conc. mM/1 NaCl*	*Urine conc. relative to blood conc.*	*Approx. urine volume as % body wt./day*
Daphnia magna	10 mgm	60	–	200*
Gammarus pulex	60 mgm	150	hyposmotic	47*
Astacus sp.	30–100 gm	220	hyposmotic	8
Potamon niloticus	15–100 gm	270	isosmotic	0·6–0·05

* Calculated values expressed in terms of % body water per day.

It seems likely therefore that there has not been a single series of stages in the physiological evolution of fresh water forms as originally suggested but more probably there have been a number of parallel routes. (1) Forms which are large and impermeable maintain a high blood concentration in brackish water and by increased powers of active uptake also sustain a high level in fresh water. These forms do not produce hyposmotic urine, e.g. *Potamon niloticus.* (2) Forms which had large but somewhat more permeable brackish water ancestors lower the blood concentration and produce hypotonic urine, e.g. Crayfish. (3) Small forms maintain a lower blood concentration both in brackish and fresh water than larger species and, in addition, at least some developed the capacity to produce hyposmotic urine prior to the colonisation of fresh water, e.g. *Gammarus* species. (4) Highly permeable forms of whatever body size have very low blood concentrations in fresh water, e.g. Molluscs and Annelids generally. They produce hypotonic urine. The urine of molluscs and annelids from dilute brackish water has not yet been studied but we should expect that it will also be hyposmotic to the blood when the animals are hypertonic to their medium.

Thermodynamic considerations suggest that the transport of ions up a small concentration gradient should be less expensive of energy than transport up larger gradients. The minimum work (W) required to move 1 gram mol of sodium chloride from a concentration A to a higher concentration B is given by the equation:

$$W = 2RT \, . \, 2{\cdot}303 \log_{10}\frac{B}{A},$$ where R is the gas constant and T is the absolute temperature.

Theoretically, therefore, it should be less expensive of energy to reabsorb ions from the excretory canal, where the concentration gradient is small, prior to the release of urine, rather than to produce isosmotic urine and then reabsorb the ions lost by taking up ions actively from the medium. In theory an animal with a blood concentration as high as that of *Eriocheir* might save up to 80–90 per cent of its energy outlay on ion transport if it produced urine isosmotic with the medium (if all the ion loss from the body were in the urine). However it is not at all certain that the production of hyposmotic urine does result in any saving of energy as some authorities believe that the energy expenditure on ion transport is independent of the concentration gradient and related only to the amount of ions transported and to the mechanism of transport. If this is so and if the mechanisms of transport are the same at the body surface and in the renal canal then the formation of hyposmotic urine can only serve the function of preventing the unnecessary loss of ions from the body, an insurance perhaps against some 'rainy day' when ions cannot be taken up at the surface as fast as they would be lost if the urine were isosmotic with the blood.

Physiological races

There are a few cases known of animals which occur in fresh water and have a much higher blood concentration than would be expected. An example of this is the isopod *Saduria* (*Mesidotea*) *entomon*, a glacial relict species which lives in fresh water in a number of Swedish lakes. The osmotic concentration of its blood has not yet been measured but the chloride concentration is about the same as that of *Eriocheir*, very much higher than we should expect in a well adapted fresh water animal of this size (circa 0·5–1 gm). *Saduria*, however, may not be a form fully adapted to life in fresh water as geological evidence suggests that it can only have colonised the Scandinavian region since the last ice age (circa 40,000 years ago). The main distribution of the species is in the brackish water arctic seas and in the Baltic, and specimens taken from the Baltic are unable to survive in fresh water even after slow acclimatisation. In this animal therefore there is an indication (a) that the specimens from the fresh water lakes constitute a distinct physiological race with greater powers of regulating their blood in dilute media than the brackish water race in the Baltic and (b) that this capacity has been recently evolved. Unlike most fresh water animals the fresh water race still retains the capacity to live in 100 per cent sea water, a factor which further emphasises its recent penetration of fresh water.

The fresh water race already shows slight morphological differences from the brackish water race and it would seem that this species gives some indication of the way in which fresh water species have evolved from brackish water ancestors, after accidental isolation from the main stock. At present the anatomical differences between the fresh water and brackish water races are only sufficient to accord them sub-specific status but there seems little doubt that if they remain isolated eventually a new species will have been formed. It is almost certain that the prawn, *Palaemonetes antennarius*, which lives in fresh water lakes in southern Europe has been derived from *Palaemonetes varians* in a similar way though in this case the anatomical differences between the forms are deemed sufficient to grant specific status to the form living in fresh water.

Summary of body fluid regulation in aquatic invertebrates

Marine invertebrates. Most marine invertebrates are isosmotic with their medium. They survive only a small dilution of the medium. Examples, *Homarus, Maia*, Echinoderms.

Brackish water invertebrates. A few forms found in the more concentrated regions of estuaries are also isosmotic with their medium but tolerate a wider range of salinities than the stenohaline marine forms. Their survival is dependent on the extent to which the body cells can adjust to dilution of the body fluids. Examples, *Arenicola* and *Mytilus.*

In very dilute media the blood must be maintained hyperosmotic to the environment even in forms which are isosmotic at higher concentrations. Example, *Nereis diversicolor.*

Highly permeable animals such as molluscs and annelids are unable to maintain a large gradient of concentration between the blood and medium. Less permeable forms such as the crustacea maintain larger gradients of concentration in media less concentrated than sea water and hence protect the cells from extreme variations in the concentration of the blood bathing them. When the blood is hyperosmotic to the medium ions are lost by diffusion and in the urine, and water is gained by osmosis. Energy has to be expended to regain the lost ions by active transport from the medium and to extrude excess water. The amount of energy required depends on the permeability of the body surface to ions and water, and on the gradient of concentration between blood and medium. Brackish-water species reduce osmotic work by having a less permeable body surface than marine forms and by gradually decreasing the concentration of the blood as the medium is diluted. Any such decrease naturally transfers the osmotic stress to the cellular level and few species can survive over the whole range from sea water to fresh water.

Some crustacea from environments liable to great fluctuations in salinity can maintain the blood hyperosmotic to dilute media and hyposmotic to more concentrated media. Examples, *Palaemonetes varians*, *Artemia salina*.

Fresh water invertebrates. These are all hyperosmotic to their medium, though again the more permeable forms such as molluscs have lower blood concentrations than the less permeable crustacea. Fresh water crustacea have lower blood concentrations than comparable forms in brackish water. Species of small body size have higher surface/volume ratios than larger forms and would have fast rates of ion and water turnover if the blood concentration were as high as that of the larger animals. Small species have lower blood concentrations than large animals.

Many freshwater species, e.g. Crayfish, *Gammarus pulex*, etc., conserve ions in the body by producing urine hyposmotic to the blood. Fresh water crabs have a high proportion of their total ion loss via the body surface and would gain little by producing hyposmotic urine. They have urine isosmotic with the blood. Examples *Eriocheir sinensis*, *Potamon niloticus*.

Some crustacea living in fresh water have higher blood concentrations than would be expected of forms well adapted to this medium. These seem to be of recent brackish water origin and are clearly related to forms confined to brackish water. Example, *Saduria entomon*.

TERRESTRIAL INVERTEBRATES

One of the principal regulatory problems of terrestrial forms is the maintenance of the body water content. There are two potential hazards (1) over-hydration by osmotic intake of water when the habitat is flooded, and (2) desiccation by evaporation when the air is dry.

Forms which have colonised land from fresh water do not

suffer from over-hydration since most already produce hyposmotic urine and hence possess a means of disposing of excess water without simultaneously depleting the body of salts. As any salt lost can only be replenished in the food, forms which have crossed the littoral zone must also develop the capacity to produce hyposmotic urine if they are to limit their salt loss. This may explain the apparent paradox that whilst many of the land crabs are able to produce dilute urine all the fresh-water crabs produce urine isosmotic with the blood.

Although over-hydration may present an occasional risk, desiccation is the main physiological hazard of life on land. Water is lost from the body by four main routes (1) by evaporation from the general body surface; (2) by evaporation from the respiratory surfaces; (3) in the urine, and (4) in the faeces.

Evaporation

The factors governing the rate of water loss from a body are complex and hence it is difficult if not impossible to predict how fast an animal will lose water in any given circumstance. However, at a first approximation, the rate of loss depends on the saturation deficit of the air, thus:

$E = a(S.D.)$, where E is the rate of evaporation, a is a constant and S.D. the saturation deficit. The saturation deficit is the vapour pressure of water in air when the air is saturated with water minus the observed vapour pressure. Alternatively this may be written $S.D. = V(100 - RH)$, where S.D. is the saturation deficit, V the vapour pressure in air saturated at a given temperature, and R.H. the relative humidity at that temperature.

The saturation deficit is a measure of the drying power of the air. Thus under ideal conditions the rate of loss of water from a body at a saturation deficit of 10 mm Hg would be twice the rate occurring at a saturation deficit of 5 mm Hg and half that at 20 mm Hg. In practice the rate of loss even from an inert body may not obey this rule. There are two main difficulties. (1) Humidity gradients build up at the surface from which evaporation is occurring and these tend to slow the rate of loss. Air currents affect these gradients and hence losses will be higher when there is a wind. This we assume sub-consciously when blowing on an inked line to dry it rapidly. (2) Evaporation tends to cool the surface being evaporated so that the vapour pressure of the liquid changes. When the water loss from animals as opposed to that from inert bodies is considered, the position is made more complicated by the fact that the permeability of the surface may itself be affected by temperature or degree of hydration. However, we are concerned here not with precise measurements of evaporation but with assessing evaporation as a hazard. This it certainly is. The difference in the vapour pressure of the body fluids of fresh water animals and that of their medium is about 1/15 mm Hg at 20°C. but the difference in vapour pressure between the body fluids and dry air at the same temperature is about 17 mm Hg. Theoretically, therefore, a fresh water animal transferred from its medium to dry air at 20° C. should lose water by evaporation at 255 times the rate at which it had been taking up water by osmosis. Hence it is apparent that animals on land must either have a reduced surface permeability or if they are highly permeable they must remain in areas of high humidity. Both alternatives have their inherent disadvantages. The body surface cannot be rendered completely watertight as the respiratory surfaces must be kept moist if oxygen is to diffuse into the body; whilst confinement to areas of high humidity has a double disadvantage. First it places severe limitations on the mode of life of terrestrial forms and secondly it is hazardous if the temperature of the humid area is liable to rise as the saturation deficit of air rises exponentially with temperature.

Water loss in the urine

Much of the water loss as urine is made necessary by the need to remove nitrogenous waste products from the body.

The major end product of nitrogenous metabolism in almost all aquatic animals is ammonia. It is eliminated from the body mainly by diffusion across the body surface as NH_3 or in exchange for cations as NH_4^+. In land animals loss via the surface is ruled out and all nitrogenous waste products must be retained in the body or removed in the urine. Ammonia is highly toxic and its concentration in the blood must be kept very low. Even though some animals can concentrate it to some degree in the urine the removal of waste nitrogen in this form by terrestrial animals would involve considerable water loss. Consequently most terrestrial animals either combine ammonia with carbon dioxide to form the relatively non-toxic substance, urea ($CO(NH_2)_2$), or excrete nitrogen as uric acid.

Urea is not so toxic as ammonia but as it is very soluble it is a convenient waste product only when a considerable amount of water can be excreted or where it can be concentrated in the urine. Uric acid is insoluble and so can be removed as a sludge with the minimum of water loss even by forms incapable of forming urine osmotically more concentrated than the blood. The main disadvantage attached to the excretion of a complex molecule such as uric acid is that its formation wastes more metabolic energy than is the case for urea.

Types of invertebrate regulation

Terrestrial invertebrates fall readily into two groups, (a) those which have permeable body surfaces and which can survive for only a short period away from areas of high humidity (annelids, gastropods, crustacea, etc.), and (b) those which have very impermeable body surfaces and which are consequently less dependent on the humidity of the environment (insects, and arachnids). In addition to these animals there are a few organisms which have a tenuous life on land in films of liquid water. These include some protozoa, planarians, nemerteans, and ostracods. Such forms, dependent as they are on the presence of liquid, are to be regarded as terrestrial by courtesy only and will not be further considered here.

(a) *forms with relatively permeable body surfaces*

Annelids. Earthworms may be regarded as physiologically fresh water forms since most of their time is spent in moist soil and they emerge only when the air is very humid. Further they will survive indefinitely in well aerated water, and, as in many fresh water animals, the excretory organs are specialised to retain salts and remove excess water. The urine is hyposmotic and cannot be made hyperosmotic to the blood and coelomic fluids. When *Lumbricus* is kept in tap water the blood has a concentration of about 90 mM/l NaCl and the urine about 15–40 mM/l. The body is ill protected against exposure to unsaturated air except in so far as the animals are exceptionally tolerant of water loss. *Lumbricus terrestris* recovers after losing up to 70 per cent of its body water. Such toleration is at best a means of surviving only short periods in an unsaturated environment; but it allows time for the worm to locate more favourable conditions. Indeed it is probable that most of the earthworms to be seen dead on the surface of the ground are killed by ultra violet light, to which they are very sensitive, or to some other cause than water loss. Earthworms have a second means of avoiding water loss. They are very sensitive to the water content of the air and if exposed to low humidities burrow deeper into the soil.

The main end product of nitrogen metabolism varies, possibly in association with the degree of hydration of the animal. Sometimes up to 90 per cent of the nitrogen loss is in the form of ammonia but up to 80 per cent can on occasions be urea.

In some earthworms the nephridia instead of opening at the body surface as in *Lumbricus*, run instead to the lumen of the gut (enteronephric nephridia). The physiological significance of this appears to be that water lost in urine production can be reabsorbed in the gut. The advantage of this as an adaptation to dry conditions can be seen by comparison of two Indian earthworms, *Eutyphaesus* and *Pheretima*. The former has exonephric nephridia whilst the main nephridia of *Pheretima* are enteronephric. Both worms occur together near the surface during the wet season, but in the dry season *Eutyphaesus* has to burrow deeply and aestivate whilst *Pheretima* can remain near the surface.

Molluscs. Though in general the terrestrial molluscs (all gastropods) are epigean rather than subterranean their bodies are little better protected against water evaporation than are the earthworms. It is possible that the copious secretion of mucus on the surface may serve as a barrier to evaporation but a convincing demonstration of this has yet to be produced. In the main their success on land is due primarily to (1) the protection against water loss afforded by the shell; (2) behavioural mechanisms limiting activity when the body water content is low and the air dry; (3) tolerance of changes in body fluid concentration, and (4) excretion of nitrogenous waste in the form of uric acid.

The concentration of the body fluids of terrestrial gastropods is in the same range as that of the earthworms but as in the latter group the concentration of individuals depends on the degree of body hydration. Thus the blood concentration of *Helix* may change from Δ 0·47° C. to Δ 0·2° C. by water uptake during a shower of rain. Slugs, which have evolved from several different groups of snails, have a similar tolerance of change in body fluid concentration. In these animals it has been found that most of the water lost when they are dehydrated is drawn from the blood and little from the cells, indicating that the cells must actively regulate their internal concentration. This may be part of the explanation of the rather low general level of blood concentration maintained by these animals since the lower the level of blood concentration the smaller is the absolute level of concentration change that must be made by cells to keep their volume constant when any given amount of water is lost from the body.

Spontaneous activity in the central nervous system is directly affected by the concentration of the blood. When the concentration is high, as it will be when the animal is short of water, activity is decreased. When the blood is more dilute activity is increased. This physiological effect can be directly correlated with the behaviour of slugs and snails since it is found that animals that are moving about always have a higher water content than those that are quiescent.

When the air and ground are dry, slugs retreat under cover to areas of high humidity and snails retire into their shells sealing the mouth with hardened mucus, calcium phosphate or intestinally processed mud. Many snails can aestivate within the shell for long periods, the classical example being the case of a specimen of *Helix desertorum* which crawled away after being attached to a tablet in the British Museum for four years.

The form in which nitrogen is excreted is closely correlated with the availability of water. During active life when it is fully hydrated and water is available for the formation of urine, *Helix* produces mainly ammonia and urea. When the animal is aestivating however, the function of the excretory organs changes. Uric acid is produced and stored within the kidney lumen. When water again becomes available the uric acid is swept out and the kidneys revert to eliminating urea and ammonia.

Arthropods. Crustacea. Three groups of crustacea have truly terrestrial members, the amphipods (hoppers), the isopods (woodlice) and decapods (hermit and brachyuran crabs). In addition there are also a number of minute forms which live in the interstitial soil water. These are mainly copepods and ostracods.

Only the isopods and amphipods have succeeded in colonising areas any distance from the sea as the crabs all have marine larval stages and must return to the sea to breed. Aquatic isopods and amphipods have suppressed the larval stages and the developing eggs are carried within a brood pouch. The eggs hatch as minute adults. In this respect these two groups were preadapted to overcome the problem of reproduction on land.

Apart from the familiar hoppers of the shore strand-line the terrestrial amphipods are limited to the deep litter of forests in the Indopacific region. Here they are common but comparative physiologists are rare and little is known about their water balance problems other than the fact that they die in a short time if not kept in a humid atmosphere.

Woodlice are found in a wide range of habitats from the extremely moist conditions at the edge of water to deserts. The water relations of the group have been extensively studied by Edney.*

The rate of transpiration loss across the body surface, though less than in the Annelida and Onycophora (Fig. 7) is consider-

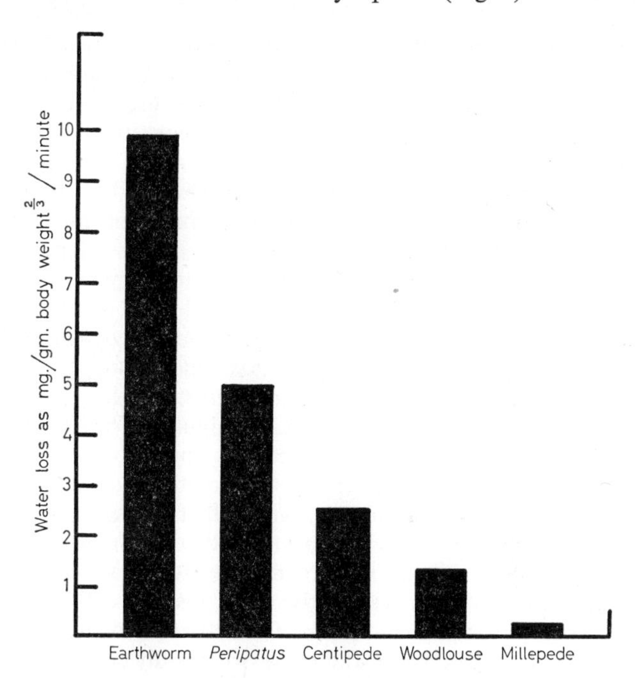

FIG. 7. Data to illustrate the differences in rates of water loss from various terrestrial invertebrates. (Values for diagram from Morrison)

able, and no woodlouse can survive in dry air for long. The permeability of the body surface is not the same in all species. Those which live in the driest places have less permeable surfaces than forms which are confined to damper situations.

* See Edney, E. B. (1954) 'Woodlice and the land habitat', *Biol. Rev.* *29*, 185–219.

In the following series the animals are associated with progressively drier situations and their permeability declines in the same order.

Ligia — *Cytilisticus* — *Oniscus* — *Porcellio* — *Armadillidium*. The transpiratory losses of *Ligia*, *Porcellio* and *Armadillidium* are related in the ratio of 1 : 0·43 : 0·38 at 30° C. and constant humidity. One of the factors limiting the extent to which evaporative losses can be decreased is the necessity of having a moist respiratory surface for gaseous exchange. Consequently, as the permeability of the general body surface is decreased the loss from the gills comes to account for a greater proportion of the total loss. In *Ligia* the loss from the gills is about 20 per cent of the total but in *Porcellio* it amounts to some 40 per cent.

Size as well as permeability affects the length of time that woodlice survive when there is a large saturation deficit in the air. Thus the large *Ligia*, although it is so much more permeable per unit area than the smaller *Armadillidium*, lives longer in dry air at 30° C. (440 minutes as against 390 minutes). The reason for this is of course the relation between the surface/volume ratio and the concentrating effect of water loss on the body fluids. For a given surface permeability the larger the body size the more slowly will the body fluids be concentrated.

Water can be taken up both by drinking and via the anus to replace water loss, and evaporative water losses are diminished by the behaviour of the animals. The activity is related to the humidity of the air. When the air is dry woodlice move about rapidly and turn infrequently. If their movements should carry them into moist air the rate of turning is increased and activity suppressed. Consequently once in a moist place they tend to remain there. Naturally this behaviour pattern would tend to trap them permanently in their hiding places if it were rigidly maintained. In fact woodlice become less sensitive to humidity at night and then move out in search of food. Even the desert woodlouse, *Hemilepistus reaumurii*, has no special physiological capacities to fit it to its environment. As night approaches in deserts the temperature often drops far enough to be below the dew point. Close to the ground therefore there is a humid or even wet layer at night. *Hemilepistus* lives in burrows by day and emerges at night.

Terrestrial isopods have a higher blood concentration than the annelids and molluscs. (Table 3).

TABLE 3. Comparison of the blood concentration of an earthworm, snail and four isopods.

Isopods	△°C.	*Annelid*	△°C.
Ligia oceanica	2·15	*Lumbricus terrestris*	0·31
Oniscus sp.	1·04		
Porcellio sp.	1·30	*Gastropod*	
Armadillidium sp.	1·18	*Helix sp.*	0·4

Data from Parry, Ramsay and Cordot.

The very high blood concentration maintained by *Ligia* is made necessary by the specialised nature of its environment. The crannies in which it lives are near the high tide mark and are periodically soaked by sea spray. Consequently, the air in them is saturated with water not at the vapour pressure of fresh water but of concentrated sea water. If *Ligia* were to have blood less concentrated than sea water it would be bound to lose water by evaporation. *Ligia*, in common with other permeable terrestrial animals, is rather tolerant of changes in the concentration of its blood. Its limits of survival are at approximately △ 1·44° C. and △ 3·48° C., the latter being almost twice the concentration of sea water.

The faeces of isopods are generally moist so that some water is lost by this route. Glands present in the rectum appear to function as water reabsorbing organs when water intake is low and then the faeces are rather more dry than usual. These rectal glands are also believed to be responsible for absorbing water taken in per anum.

Considering that they are terrestrial the isopods secrete an unusually large proportion of their waste nitrogen in the form of ammonia. Up to 70 or 80 per cent of the non-protein nitrogen lost by *Ligia* is as ammonia and even in the more fully terrestrial forms some 50 to 60 per cent of the total is in this form. Uric acid accounts for much of the remainder but the isopods are curious in that uric acid is also retained in the body. The species which can occur in the driest habitats tend to retain most uric acid, *Armadillidium* having more than other terrestrial forms. However, this relationship may be fortuitous as the body of the fresh water species *Asellus aquaticus* contains five times as much uric acid as *Armadillidium*. Although a high proportion of the waste is as ammonia the rate of output of nitrogen by terrestrial forms is lower than that of aquatic species and it may be that the terrestrial isopods actually decrease nitrogen metabolism.

Crabs. The crabs of warm coasts form an interesting series showing increased behavioural and physiological adaptations to terrestrial conditions.

The rock crab, *Pachygrapsus crassipes*, is a typical mid- to high-tide inhabitant of rocky shores. It regulates its blood hyper or hyposmotic to the medium according to the salinity, but the excretory organ contributes to the regulation of the blood only by the differential excretion of ions not by producing hyperosmotic or hyposmotic urine. If the animal's blood concentration is altered outside the usual range and it is then given a choice of media it tends to select 100 per cent sea water rather than the salinity which would most rapidly bring the concentration back to normal.

Various species of the fiddler crab, *Uca*, make burrows just below or above high tide mark. The burrow extends down sufficiently far to reach moist sand so when in its burrow the crab is in a nearly saturated atmosphere. Unlike *Ligia*, however, *Uca* maintains its blood hyposmotic to sea water. Consequently, it will tend to lose water by evaporation when it is in its burrows and by osmosis when submerged by sea water. A mechanism must therefore be present which can dispose of salt more rapidly than water. When *Uca* is in media more concentrated than its blood it drinks and produces urine hyperosmotic to the blood. This conserves water while removing excess salts. The outward secretion of salt at the gills may also play a part in the regulation of the blood concentration.

The Ghost crab, *Ocypode*, makes burrows above the high tide mark but tends to visit the sea at night. Like *Uca* it can produce hyperosmotic urine when in media more concentrated than the blood. When it is out of water however it conserves water by suppressing the initial formation of urine. This is shown by the fact that when the carbohydrate inulin, which is not metabolised, is injected into the blood it stays there. Suppression of urine formation appears to be a successful mechanism covering temporary periods out of water, since, by contrast, the mangrove crab, *Goniopsis*, which clears inulin from its blood at a constant rate in different media, dies within 24 hours out of water. The retention of metabolic wastes in the blood renders mere suppression of primary urine formation unsatisfactory for more fully terrestrial crabs such as *Gecarcinus*. This land crab removes injected inulin from its blood, presumably by the formation of primary urine, but there is no large release of urine from the body. Indeed, there is evidence that most of the water and salts in the primary urine must be withdrawn from the renal canal. This process serves to conserve the valuable salts and water while concentrating waste matter. However, the capacity to reabsorb water and salts in the excretory organ is not a new mechanism developed in this animal but merely a marked extension of a similar but limited reabsorption of salts and water which occurs when the common shore crab, *Carcinus*, is kept out of water (Fig. 8).

Gecarcinus, like the semi-terrestrial crabs of the shore, restricts evaporative water loss by spending much of its time in burrows. The hermit crab *Coenobita* on the other hand does not burrow but like many of its marine relatives inhabits mollusc shells. Some 90 per cent of its time is spent out of water and it is therefore subject to considerable evaporative losses. In compensation for this it shows, like *Ligia*, a very considerable tolerance of changes in the concentration of the blood. It survives variations in the range 83–220 per cent of the normal level. In addition, when the blood concentration is varied from the normal, *Coenobita* visits media of the appropriate salinity to correct body fluid level, an advance over the behaviour of the less terrestrial *Pachygrapsus* in similar circumstances.

Myriapoda. Millepedes and centipedes like the crustacea lack a wax layer in the cuticle and tend to remain in areas of high humidity. Little detailed knowledge is available however as to their water and salt regulation.

(b) *forms with relatively impermeable body surfaces*

Insects. Insects are the most successful land invertebrates and, in terms of both numbers of species and individuals, the most successful of all animal groups. They have colonised almost

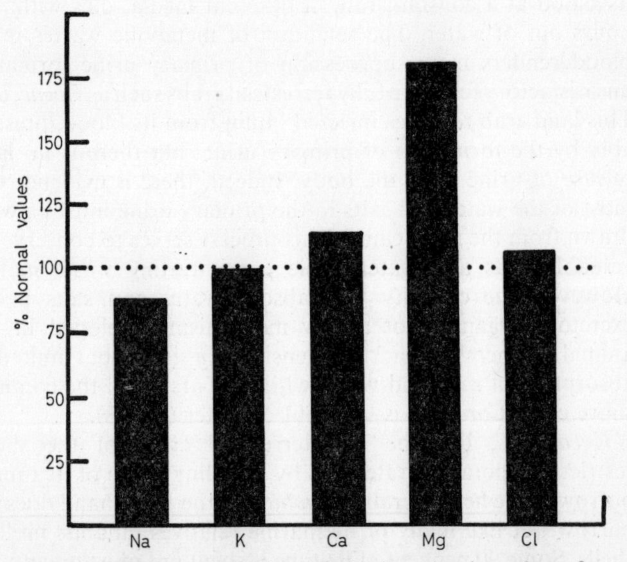

FIG. 8. The concentration of the ions in the urine of *Carcinus* when the animals have been out of water for 72 hours. (Data for diagram from Riegel & Lockwood)

every conceivable ecological niche on land and in fresh water. Only in the sea are they uncommon. Much of their success can be attributed to their exceptional capacity for restricting water turnover. Water loss in the terrestrial forms is limited by:

(1) A highly impermeable body surface.
(2) The withdrawal within the body of the moist surfaces where respiratory exchange occurs.
(3) The production of urine more concentrated than the blood.
(4) Elimination of waste nitrogen as uric acid.

The principal barrier to water movement in the insect cuticle is located in the outermost region, the epicuticle. The effective water proofing agent is a layer of grease or wax, the thickness of which in most insects is probably no more than 0·1–0·3 μ (0·0001–0·0003 mm). Any factor such as abrasion, fat solvents or high temperature which breaks down this wax layer brings about a considerable change in the permeability of the cuticle and complete removal of the wax may increase the rate of water loss by up to 100 to 300 times. The impermeability of the layer of wax is greater than would be expected of a layer of this thickness if all the wax molecules were randomly orientated, and it is therefore probable that at least one layer must be present in which the wax molecules are aligned so as to present an almost impenetrable barrier to the passage of water molecules (fig. 9a). Beament, studying the effects of variation in temperature, has confirmed this assumption. As the temperature is raised there is at first little change in the rate of water movement across the cuticle. Then at a critical temperature, the transition temperature, there is a sudden increase in the rate of water loss. If the temperature is then lowered again the water loss decreases, but only gradually (Fig. 9 d). This result is taken to indicate that at the transition temperature the ordered wax layer is disrupted so that the molecules become randomly orientated (Fig. 9 b). When the temperature is lowered again below the transition point the molecules cannot be reorientated in a species which has a hard wax (long chain molecules) and so the rate of loss remains high. The cockroach is covered by a layer of grease (short chain molecules) rather than a hard wax, and in this case the orientated

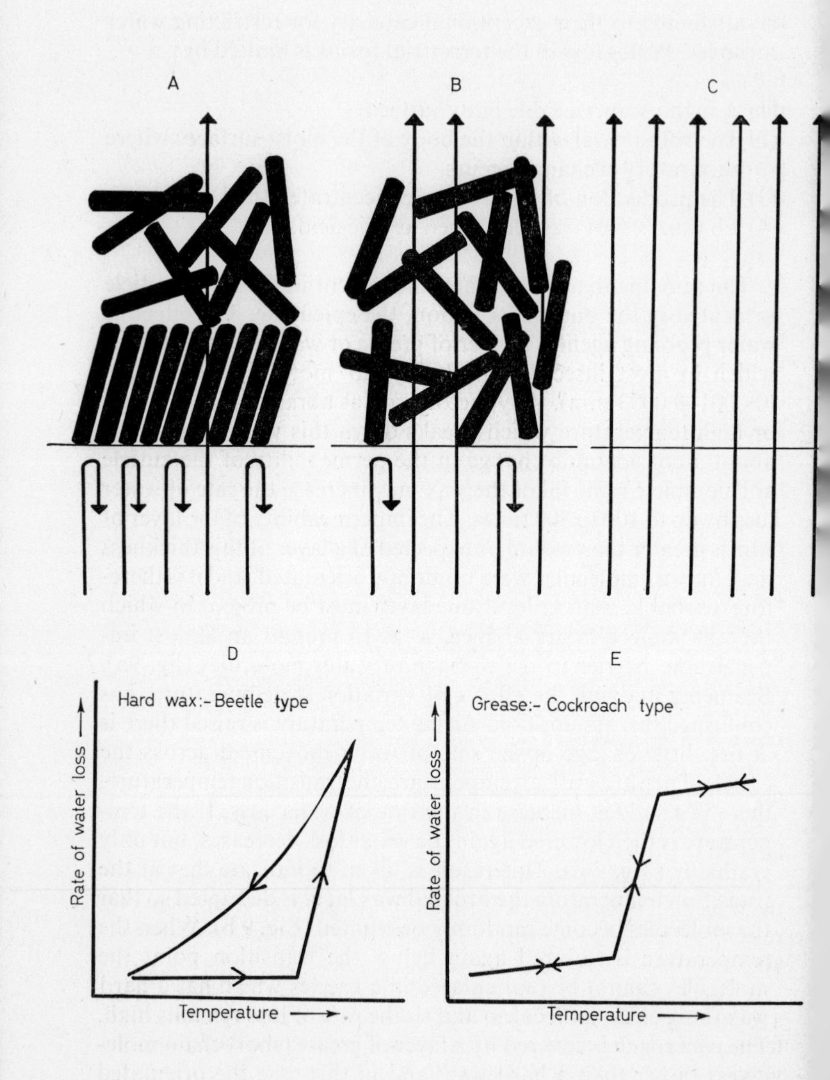
A
B
C
D
E
Hard wax:- Beetle type
Rate of water loss
Temperature
Grease:- Cockroach type
Rate of water loss
Temperature

layer can apparently be re-formed again on re-cooling (Fig. 9 e).

The temperature at which transition occurs is related to the normal environment of the species. Insects living in hot dry places tend to have high transition temperatures, those from aquatic habitats often have waxes with rather low transition temperatures. Thus the transition temperature for the cockroach is at about 33° C. whilst that of the water beetle, *Dytiscus*, is only 24° C. The wax layer of the terrestrial forms serves to prevent water loss by evaporation whilst that of the aquatic species prevents the osmotic uptake of water. If the temperature of the water is raised to about 26° C. the water proofing of *Dytiscus* breaks down and the animal, taking up water by osmosis, becomes water-logged and dies. Beetles such as *Agabus* which can live in warmer waters have transition temperatures at near the cockroach level.

A novel feature of terrestrial insects is that some of them can take up water from the air when the relative humidity is high. The only other animals known to be able to do this are a few mites. The flour beetle, *Tenebrio*, absorbs water from air with a

---

FIG. 9. (*Opposite*) Diagrammatic representation of the arrangement of the molecules in an insect cuticular wax layer and of the effect of temperature change on water loss.

(a) Normal situation in which an ordered layer of grease or wax molecules forms an effective barrier preventing water loss. Additional wax molecules are randomly orientated.

(b) Heating the cuticle to a temperature above the transition point disrupts the ordered layer and greatly increases water loss.

(c) Removal of all grease molecules with a solvent results in a large water loss.

(d) The ordered layer cannot be re-formed after being heated above the transition temperature in forms with hard waxes. Thus even on re-cooling below the transition temperature the water loss remains above normal.

(e) The ordered layer is re-formed on cooling in species which have a grease layer. In such forms the water loss is constant at a given temperature both before and after being heated above the transition temperature. (Partly after Beament)

relative humidity of 90 per cent and the prepupal stage of the flea *Xenopsilla* can even take water from air with a relative humidity of 58 per cent. Such water uptake is not the result of the vapour pressure of the blood being less than that of the air. Even at a relative humidity of 90 per cent the body fluids would have to be some 15 times their actual concentration for this to be possible.

The tracheal respiration of insects reduces the water loss associated with gaseous exchange, but nevertheless, loss of water from the tracheae constitutes a considerable proportion of the total loss. In the case of a grasshopper about 70 per cent of the water lost is estimated to escape from the spiracles. Loss from the tracheae is minimised as the tracheal apertures are sealed by spiracles except during respiratory movements. Evaporative losses are increased by any factor which raises the rate of respiration or causes the spiracles to be held open for longer periods. Thus exposure to a high concentration of CO_2 which causes the spiracles to be kept open greatly increases the water loss from the blood sucking bug, *Rhodnius*. Similarly a rise in temperature, by raising the metabolic rate and hence respiratory rate, also results in increased water loss. As in other terrestrial forms behaviour may supplement physiological control of water loss. Locusts adopt a behavioural means of limiting the rise in body temperature and hence water loss during the heat of the day. At day-break they are randomly oriented but as the sun moves higher they rotate the body so as to ensure that the minimum surface area is exposed to direct illumination.

Toleration of dehydration is lower in many insects than in other terrestrial invertebrates, presumably because the efficiency of water retention is such that in normal circumstances the body water content rarely falls. *Popillia* larvae, *Tenebrio* larvae and unfed *Rhodnius*, to quote three examples of animals with very different rates of water availability in their habitat, all die when their water content has fallen by about 20–25 per cent. The insects can however provide an exception to practically every generalisation one may care to make, and Hinton has found that the larva of the chironomid midge, *Polypedilum vanderplanki*, can be revived after it has been almost completely dehydrated. Not only does it survive the loss of water, but in this condition it will withstand being heated to 68·5° C. for 11 hours. It regains normal activity on being hydrated. After being heated to 200° C. for five minutes a few again recover normal activity on being rehydrated but some irreversible change is caused by this treatment as the animals die after about 20 hours. The exceptional tolerance of water loss shown by this animal is necessary in its habitat. The larvae live in small water-filled holes in rocks in Nigeria and these are liable to dry out before the larval development is complete. The capacity to survive in a dehydrated state between the rains must therefore contribute to the viability of the species.

Some termites from arid environments adopt an ingenious means of obtaining water. In common with other insects from dry places they produce faeces which have a very low water content. The faeces being hygroscopic take up water when the dew falls at night. The termites then consume the faeces and reduce the water content once more thus gaining the excess water.

In most animals the major part of the osmotic concentration of the blood is accounted for by sodium and chloride. In insects however both the total concentration and the ionic composition of the blood shows great variety from one species to another. In so far as generalisations are possible it may be said that the Apterygota (flightless forms such as the Thysanura, Collembola, etc.) and the Exopterygota (Orthoptera, Odonata, etc.) have blood which is approximately equivalent to that of non-insect groups in having a high sodium concentration, low potassium concentration and with a large proportion of the total osmotic pressure accounted for by inorganic ions. In the Endopterygota great variety is found, the blood composition, though genetically fixed in any one species, showing a relation to the diet. Blood-sucking forms and parasites such as the horse fly, *Gastrophilus intestinalis*, have high sodium and low potassium concentration in the blood but the larvae of *Drosophila melanogaster*, which feeds chiefly on the yeasts associated with rotting fruit, has almost equal concentrations of sodium and potassium in the blood. In this animal a considerable proportion of the blood osmotic pressure is accounted for by organic molecules, particularly amino-acids. *Drosophila* larvae have remarkable powers of maintaining the ion ratios of the blood. A race has been bred which will develop from the egg to the adult on food to which 70 grams of NaCl per kilogram has been added. These

TABLE 4. The cation concentration of the haemolymph of various insects.

	Na *mEquivs/*1	*K* *mEquivs/*1	*Ca* *mEquivs/*1	*Mg* *mEquivs/*1
Apterygota				
Petrobius maritimus A	208	6	–	–
Exopterygota				
Periplaneta americana A	156	8	4	5
Endopterygota				
Gastrophilus intestinalis L	206	13	7	38
Bombyx mori L	3	42	25	81

A Adult
L Larva

animals will also grow on media containing 70 grams of KCl per kilogram food. These high concentrations of inorganic ions make only small differences to the ionic composition of the blood. It is difficult to imagine an animal from any other group but the Diptera tolerating such diets.

The most extreme variation from the typical ionic composition of blood is found in Lepidopteran larvae such as the silk-worm, *Bombyx mori*. These have vanishingly small amounts of sodium in the blood and high levels of potassium, magnesium and amino-acids (Table 4).

Arachnids. Water loss in the various terrestrial groups of arachnids (scorpions, mites, spiders, etc.) is restricted as in the insects by a wax layer in the cuticle. In the spiders that have been investigated the critical temperature is about 30° C. but doubtless forms such as the scorpions and spiders from hot regions have higher critical temperatures. Arachnids like insects eliminate waste nitrogen in an insoluble form but the substance excreted is guanine not uric acid.

Summary of body fluid regulation in terrestrial invertebrates

A major physiological problem in the invasion of terrestrial environments has been the regulation of the body water balance. Forms with permeable surfaces face two hazards, the uptake of water by osmosis when the habitat is inundated with 'distilled' water (rain) and water loss by evaporation when the air has a relative humidity less than 100 per cent.

Terrestrial invertebrates can be divided into two groups: (a) forms with fairly permeable body surfaces, annelids, molluscs, crustacea, myriapods, etc. (the rate of water loss by evaporation varies widely in these groups Fig. 7), and (b) forms with very impermeable body surfaces the insects and arachnids. An insect such as *Rhodnius* loses water at only about one-tenth the rate of the millepede shown in Fig. 7 in comparable conditions.

On land, inorganic ions are available only in the food and hence must be conserved in the body. Most terrestrial forms are able to conserve ions by forming urine hyposmotic to the blood. In addition some crustacea and insects can also produce urine hyperosmotic to the blood in conditions where water is at a premium.

Survival in dry environments depends on both behavioural and physiological mechanisms.

Permeable forms. Behavioural mechanisms. Permeable terrestrial forms such as the woodlice are sensitive to the humidity of their environment and have developed behavioural responses that restrict them to local humid niches when the air is dry. Earthworms burrow more deeply when the surface soil dries out. The activity of molluscs is related to the degree of hydration of the body. Partial dehydration results in a decrease in activity, rehydration promotes activity and emergence from cover. This behaviour can be related to the effect of changes in the blood concentration on the spontaneous activity of the central nervous system.

Physiological mechanisms. Many terrestrial forms, particularly the more permeable forms such as the annelids and molluscs, show great tolerance of water loss. There is evidence which suggests that extracellular fluid loss is greater than that from the cells when the animals lose water, which indicates that the cells may adjust to changes in body fluid concentration as in brackish water forms.

Nitrogenous waste is excreted in the form of urea (annelids) or uric acid (onycophora, crustacea and molluscs) when water intake is low. Ammonia and urea may be released in larger amounts when water intake is greater.

Less permeable forms. The insects and arachnids have a layer of wax in the cuticle which restricts water loss. In the insects one layer of the wax molecules is aligned so as to present an almost impenetrable barrier to water molecules. There is consequently a great increase in the rate of water loss through the cuticle at the temperature which just disrupts this barrier. Insects with hard waxes (long chain molecules) cannot re-form the organised layer once it has been disrupted. Insects such as the cockroach which have a grease (short-chain molecules) can re-form the layer when cooled again below the transition temperature.

Most of the water loss from insects is via the spiracles. Loss by this route is dependent on metabolic rate and hence on activity and body temperature. Locusts orientate to the sun in order to expose the minimum surface to radiant heat. Water retention is so efficient in most insects that they are rarely exposed to dehydration. They are intolerant of severe water loss.

The main excretory end product of insects is uric acid, that of arachnids is guanine. The faeces are dry in forms from arid environments. Some vegetarian Holometabola are unique in the animal kingdom in that the blood has a low sodium content and high concentrations of potassium and magnesium.

3

Vertebrate Body Fluid Regulation

All vertebrates, except the hagfish, bear the physiological impress of the evolution of early vertebrates in fresh water. Irrespective of the conditions in which they now live, they have blood salt concentrations equivalent to about one quarter to one half sea water. The maintenance of this level involves, as in the case of invertebrates, a variety of different forms of regulation according to the nature of the environment.

PRIMARILY AQUATIC FORMS

Hagfish

The hagfishes, of which *Myxine* is an example, have body fluids which are approximately isosmotic with sea water. Furthermore, they are unique among the few vertebrates which are isosmotic in that the major part of the osmotic pressure of the extracellular fluids is accounted for by inorganic ions (Fig. 10). There is thus no great tendency for water or ion movements across the body wall and their osmotic regulation must be rather similar to that of the marine invertebrates.

Fresh water teleosts and lampreys

The hagfishes are marine but the other group of cyclostomes, the lampreys, contains many species which spend part of their lives in fresh or brackish water. In fresh water the lampreys have a blood concentration equivalent to about one quarter that of sea water though the level rises to about one third sea water in more saline media. Fresh water teleosts also have a blood concentration which lies in about the same range, $\triangle$ 0·45° C.

Tinca to △ 0·64° C. *Salmo*. Both lampreys and teleosts are thus very hypertonic to their medium when in fresh water, and they take up water by osmosis. This water entry is mainly via the gills in the teleosts but probably occurs more generally over the scaleless surface of the lampreys. Water equivalent to as much as 30 per cent of the body weight may be taken in daily and this is eliminated by the copious production of urine. The urine is hyposmotic to the blood and may be as dilute as △ 0·04 °C. but it nevertheless constitutes a drain on the body salt reserves greater than the normal rate of salt uptake in the food. Loss of salts is made good by active uptake of ions through the gills,

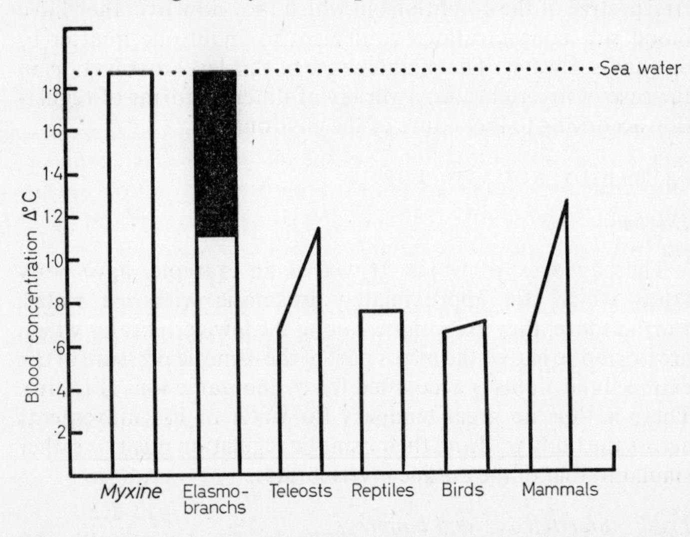

FIG. 10. The approximate blood concentration in various marine vertebrates. Only the elasmobranchs and *Myxine* have blood isosmotic with the medium. Note the considerable contribution of urea (black) to the total concentration of the elasmobranch blood.

though there is some evidence that very young fish may take up ions more generally over the surface. Fresh water fish do not normally drink their medium.

Marine teleosts

The blood concentration of marine teleosts though a little higher than that of the fresh water forms is still only about half that of sea water and hence these animals are very hyposmotic to their medium. The actual level varies somewhat from species to species. Thus the concentration of the blood of mackerel is equivalent to $\triangle$ 1·13° C. whereas that of the stickleback, *Gasterosteus* is only about $\triangle$ 0·65° C. Sea water freezes at –1·86° C.

The regulation of the blood hyposmotic to the medium results in osmotic water loss through the gills and to a lesser extent through the integument as a whole. Water loss can be replaced by drinking. However, as water is apparently absorbed from the gut by osmosis it can only be taken into the body provided that a major part of the ions in sea water are also absorbed. To this end sodium chloride is actively taken up from the gut. Now uptake of sodium chloride would raise the concentration of the blood were not the salt continuously removed. Removal of the salt occurs through the gills, and Keys and Willmer suggested that certain very large cells present in the gills of teleosts are responsible for its elimination.

Some support for this theory comes from study of the lamprey, *Lampetra fluviatilis* which, when it is in saline water, has cells similar to those of the teleost. During its breeding migrations into fresh water these cells on the gills decrease in number and at the same time the capacity of the animals to regulate the blood hyposmotic to 50 per cent sea water declines markedly. It should be stressed however that there is still no certainty that these cells are the only ones involved in salt secretion, and in recent years their functional significance has been repeatedly questioned.

The kidneys of marine teleosts play no part in the osmotic regulation of the body fluids as they are unable to form urine hyperosmotic to the blood. The kidneys do however have an important role in the maintenance of the ionic composition of the blood. Usually, little of the magnesium, calcium, and sulphate present in the sea water which is drunk is absorbed from the gut and most passes out in the faeces. However, some of these ions are taken in, particularly when, as in the case of fish in poor condition, the water turnover is greater than normal. Magnesium, calcium, and sulphate are then removed in the urine.

Fish which migrate from sea water to fresh cannot instantaneously switch from hyposmotic to hyperosmotic regulation. They must spend some time in brackish water to become fully acclimatised to dilute media. Elvers, when they migrate from the marine breeding grounds, remain in river estuaries for a time before moving into fresh water. Salmon are often more precipitous in their run into fresh water and are not infrequently incapacitated for a time on their first entry into this medium. They then tend to lie up in the first pool until their regulation is normal, a fact which makes them an easy prey to the intelligent poacher.

Marine teleosts living in arctic waters are in danger of being frozen solid when the temperature of their medium falls below the freezing point of their own body fluids. Forms which live in fairly deep water can be supercooled without their body fluids freezing. Surface living species cannot afford to be supercooled, as contact with floating ice would at once result in their body fluids crystallizing out, and they do not survive being frozen solid. Instead of allowing passive supercooling many of the surface species increase the concentration of the blood so that the body fluids have a lower freezing point. The nature of the anti-freeze substance used to concentrate the blood has yet to be discovered but it does not appear to be inorganic ions, urea or sugars. Another means of preventing the body fluids freezing seems to be associated with a raised colloidal content of the blood. It is thought that it is possible for colloid to form a protective coating round an ice crystal so that even if the blood is seeded with ice when supercooled it will not at once freeze throughout.

Elasmobranchs

Elasmobranch blood contains about the same amount of salt as that of the marine teleosts, but the former group of fish retain the excretory waste produce urea in the blood to such an extent that the total osmotic pressure of the body fluids exceeds that of sea water (Fig. 10). The maintenance of water balance thus presents no problem to these animals as there is no tendency for water to be lost across the body surface.

Urea penetrates readily into cells so in solving the osmotic problem in this way the elasmobranchs must also have solved the biochemical side effects of the presence of large amounts of urea in the tissues. Urea in high concentrations disrupts the hydrogen bonds linking the peptide groups of neighbouring protein molecules. It therefore tends to alter the solubility and configuration of proteins and hence their properties. Stabilisation of proteins against this effect must require modification and it is presumably because of these modifications that forms such as the saw-fish *Pristis* which can move into brackish or fresh water do not then completely eliminate urea from the blood even though it aggravates the osmotic entry of water. It is said that the heart of elasmobranchs cannot beat in the absence of urea.

Urea is retained by the kidneys, and the gills, unlike those of most fish, are impermeable to this substance. Another unusual feature is that all the body tissues seem to contribute to urea production whereas, in most vertebrates, this function is confined to the liver.

Though the blood of elasmobranchs is slightly hyperosmotic to their medium there is still a considerable gradient of salts between the blood and sea water. Further, any invertebrates eaten will contain salts. The absence of the need to drink the medium to regain lost water does not therefore absolve these fish from the necessity of eliminating salt from the body, though it is true that the amount to be removed is very much smaller than in the case of the teleosts. The elasmobranch renal tubules are exceptionally long but this seems to be associated with the re-absorption of urea and not with the concentration of salt, as their kidneys are unable to secrete urine with a chloride concentration greater than that in the blood. The gills apparently do not participate in salt removal, and Keys and Willmer failed to find 'chloride secretory cells' on the gills of the dogfish. It was not however until 1960 that it was discovered that it is the enigmatic rectal gland which is responsible for salt excretion. The rectal gland is confined to elasmobranchs and takes various forms. In the dogfish it is a thick-walled finger-like expansion of the gut opening at the base of the spiral valve into the top of the rectum. It secretes a fluid isosmotic with the blood, but this fluid differs from the blood in that it is almost pure NaCl and contains only traces of urea. The rate of secretion is variable but is probably adequate to account for the removal of all excess salt from the body.

Protopterus

The retention of urea is not limited to elasmobranchs. The main end products of nitrogen metabolism in fish are ammonia and trimethylamine oxide but a small proportion of urea is also formed. Trimethylamine oxide is removed from the body in the urine but as long as the gills are in contact with water ammonia can escape by diffusion. When a fish is out of water ammonia might be expected to accumulate in the blood, but this does not happen in the lung fish, *Protopterus*, which can aestivate in dry mud for many months when its fresh water medium dries up. This animal ceases to form ammonia in large quantities when aestivating and instead forms and accumulates urea. As much as 2 to 4 per cent urea may be present in the body fluids after several months of aestivation, an amount close to the elasmobranch level. Indeed water shortage, which stimulates urea production in *Protopterus*, may have been the biochemical stimulus which was initially responsible for the increase in the urea level of elasmobranchs when they moved from fresh water to the sea. We shall see shortly that the only amphibian to inhabit strongly saline waters adopts the same mechanism.

PRIMARILY TERRESTRIAL FORMS

Amphibia

Amphibia, though far from being the most successful of vertebrates, have individual species which inhabit almost every environment from fresh water to deserts. The clawed toad *Xenopus* and the mud puppy, *Necturus* are examples of forms that are normally fully aquatic, though *Xenopus* can aestivate in mud if its medium dries up. The frog genus *Rana* contains species which show various degrees of association with water from the almost fully aquatic *R. grylio* to species such as *R. terrestris* that may spend much of their time out of water. Toad species such as *Bufo terrestris* live in still drier conditions whilst *Scaphiopus* and a number of other toads may live in quite arid regions. Not even these last forms are completely emancipated from water however as they have aquatic larvae. No amphibian species have adults which live in sea water but *R. pipiens*, *R. esculenta* and *R. cancrivora* may all occur in brackish water. The range of habitat and the physiological problems involved in the maintenance of the blood concentration are thus even more varied in amphibians than in fish.

In the few species studied the blood has a freezing point depression of about 0·45° C. and hence in the case of species living in fresh water there is a considerable gradient of concentration between the blood and the medium. The skin is scaleless and fairly permeable so water is taken up by osmosis. Injections of hormones from the pituitary gland increase the permeability of the skin to water and it is possible that the permeability may be actively controlled in life. Water intake can be considerable, up to 25–30 per cent of the body weight per day in *Xenopus* and *Rana*. This water is eliminated as urine and, as in the fresh water fish, the urine is very dilute. Nevertheless, the urine contains more salts than are usually available in the food and their loss has to be made good. Gills are absent in adult amphibia other than the neotenic species and active uptake of ions or at least of Na, Cl and K takes place through the whole body surface.

The principal end-product of nitrogen metabolism in aquatic species is ammonia though some urea is also formed. (Table 5). Most of this ammonia passes out in the urine. When *Xenopus* is deprived of water intake by placing the animal in moist air or isosmotic salt solutions, adequate water is no longer available to remove waste nitrogen from the body as fast as it is formed. Thus when the animal is aestivating the problem of the disposal of waste nitrogen would be expected to become particularly acute, especially as starving amphibia tend to metabolise mainly protein and hence increase their production of waste nitrogen. In fact poisoning by the accumulation of ammonia in the body is avoided by the cessation of the kidney's production of this substance when the water intake falls. Instead the liver forms increasing amounts of urea. As urea cannot be concentrated in the urine of *Xenopus* any excess is allowed to accumulate in the body until water becomes available for its removal. The parallel between this response to water shortage and that of the dipnoan, *Protopterus* is striking.

Amphibia never maintain the blood hyposmotic to their medium. When they are placed in hyperosmotic salt solutions the normal direction of the osmotic gradient across the skin is reversed and so is the direction of water flow. The body fluids are concentrated by dehydration until such time as medium and body fluids are isotonic. Aquatic amphibians do not in general tolerate an increase in the concentration of their blood and most species die if placed in media more concentrated than that of their normal blood. In this they resemble most fresh water invertebrates. The maximum concentration of medium tolerated for any length of time is approximately one third sea water. The crab-eating frog *Rana cancrivora* mentioned above is unusual in that it tolerates concentrations considerably higher than this. In the laboratory it will survive in 80 per cent sea water if acclimatised slowly to gradually increasing concentrations. The tadpoles of this species are even more remarkable than the adults as they have been found in ditches containing full strength sea water. Part of the secret of tolerance of high salinity by *R. cancrivora* lies in its retention of urea when the medium concentration is raised. The blood concentration is maintained hyperosmotic to the medium at all levels by successive increments of urea. *R. cancrivora* has therefore adopted the change in nitrogen excretion associated with restricted water intake which occurs in *Xenopus*, to subserve an osmoregulatory function. The parallel with the elasmobranchs is obvious.

Examination of Fig. 11 shows how the regulation of *Rana cancrivora* differs from that of a typical frog such as *R. tigerina*. When the concentration of the medium bathing *R. tigerina* is raised the plasma concentration rises. Urine volume probably falls as the concentration gradient between blood and medium declines but chloride in the plasma is kept fairly constant by a

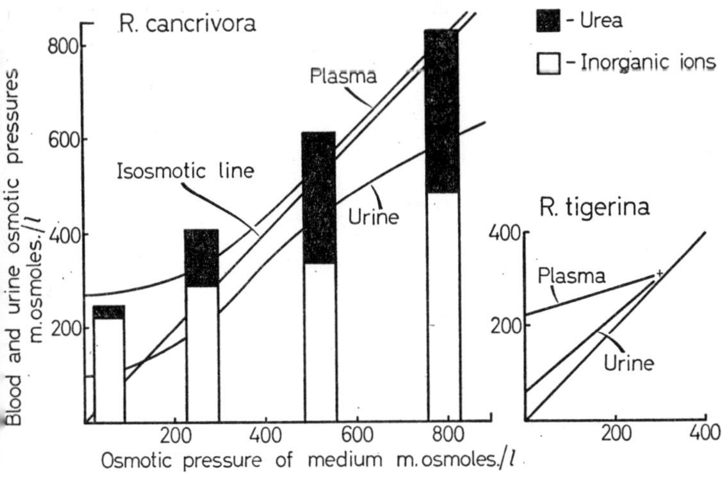

FIG. 11. Comparison of the osmotic regulation of *Rana cancrivora* and *R. tigerina* in saline media. Note the successive increments in the urea present in the blood of *R. cancrivora* when the concentration of the medium is raised. (After Gordon, Schmidt-Nielsen and Kelly)

big increase in the strength of the urine. At a concentration of the medium of about 280 m. osmoles the urine chloride is equivalent to that of the blood and the total concentration of the blood is only a little above that of the medium. Any further increase in the medium is therefore bound to be followed by a parallel increase in blood ion concentration. This is insupportable and 280 m. osmoles is the maximum level *R. tigerina* will tolerate.

R. cancrivora is more tolerant of raised blood ion levels than most amphibia but it is undoubtedly the retention of urea which enables this animal to tolerate such high salinities without the accompanying rise in ion levels and restriction of urine volume which prove fatal to *R. tigerina*.

Urea has to be retained in the body of *Rana cancrivora* and when the animal is in media more concentrated than fresh water, the level of this substance in the urine is less than that in the blood. Terrestrial amphibia also have urea as the main end product of their nitrogen metabolism (Table 5), but their

TABLE 5.

		ammonia as % NH_3 plus urea
Pipa pipa	aquatic	92·5
Xenopus laevis	aquatic but able to aestivate	62·2
Rana esculenta	semi-aquatic	9·4
Bufo bufo	terrestrial (dry)	5·7

Data from Cragg, Balinsky and Baldwin.

problem is the reverse of that of *R. cancrivora*, namely to remove as much urea as possible in a limited volume of urine. Urea is concentrated in the renal tubules of both *Rana sp.* and *Bufo calamita* thus increasing the amount eliminated. The urea concentration cannot be very large, however, as amphibia, like fish, are unable to produce urine hyperosmotic to the blood.

Evaporation is a major hazard to terrestrial amphibia as loss of water through the skin can be extremely rapid if the air is dry and warm. Four mechanisms contribute to the survival of terrestrial forms:

(1) Reduction of the permeability of the skin.
(2) Modifications of behaviour tending to restrict the animals to places of high humidity.
(3) Tolerance of water loss.
(4) Rapid absorption of water when it is available.

At room temperature and humidity a frog loses about 25 per cent of its body weight by evaporation in 10 hours. At higher temperatures the rates of loss are increased and may be quite spectacular. Thus Gordon finds that *R. cancrivora* loses about 20 per cent of its body weight in an hour at 31° C. and 55 per cent relative humidity. Rates of loss similar to that shown by *R. cancrivora* would render life in areas of low humidity and high temperature quite impracticable and some species from drier areas have greatly decreased the permeability of the skin. Figures strictly comparable with those for *R. cancrivora* are not yet available but the value given by Chew and Dammann for the fossorial toad *Scaphiopus couchi* kept at a slightly lower temperature and relative humidity (27° C. and 3·3 mg. water/litre air) is strikingly lower than that for the semiterrestrial frog. The toad loses water equivalent to only 0·6 per cent of its body weight per hour.

Many toads live underground emerging only at night when the temperature is low and the air fairly humid. Some dig their own burrows and *Scaphiopus* receives its nickname, the spade foot toad, from the modifications to the rear feet which facilitate digging. In the burrows the humidity may be expected to be fairly high; but, despite this, the animals may at times be exposed to dehydration. Tolerance of water loss is related to the potential danger of dehydration and hence to the general dryness of the normal habitat. (Table 6).

TABLE 6. Loss of water as % required to kill

Scaphiopus holbrookii	60·2	fossorial
Bufo terrestris	54·9	terrestrial
Rana pipiens	44·9	terrestrio-aquatic
Rana grylio	38·0	aquatic

Data from Thorson and Svihla.

Evaporation to the extent tolerated by *Scaphiopus* would be expected to more than double the body fluid concentration. It is not known if or how the animal compensates for this.

In the genus *Neobatrachus* there are species from a number of habitats of varying dryness but these animals all die when the body water content has been reduced by about 40–45 per cent.

In this genus however there is a marked difference between different species in the rate at which water is taken up through the skin. Species from the drier environments have faster rates of uptake.

This is curious and presents something of a physiological puzzle. It would be expected that if the rate of uptake of fluid is fast from liquid, then the rate of evaporative loss would also be rapid when the air is dry. It is not known how desert *Neobatrachus* species avoid this hazard but either the permeability or the vascular supply to the skin must be capable of variation.

In some toads water can be withdrawn by osmosis from the bladder into the blood if the permeability of the bladder wall is increased by a hormone from the pituitary gland. The bladder is thus potentially an emergency water store and the desert frog *Chiroleptes platycephalus* appears to use it for this purpose. The bladder volume is large in some toads, up to 30 per cent of the body weight in *Bufo cognatus*, thus increasing the functional capacity as a water store.

Reptiles

Species of reptile live in fresh water, on land and in the sea. The blood concentration of some fresh water species tends to be lower than that of terrestrial forms and these in turn lower than marine species. The range of concentration is however similar to that found in other primarily terrestrial groups. (Table 7).

TABLE 7.

		Blood Conc. Δ °C	*Normal Habitat*
Emys europea	(terrapin)	0·44	fresh water
Testudo graeca	(tortoise)	0·60	terrestrial
Trachysaurus	(lizard)	0·65	terrestrial
Caretta caretta	(turtle)	0·76	marine

Fresh water reptiles. The skin of reptiles is heavily keratinised and effectively impermeable to water and salts. The kidney is able to conserve salts by the production of a dilute urine. Fresh water reptiles cannot therefore be regarded as having any particular problem in the regulation of their water and salt balance.

Terrestrial reptiles. These are more fully emancipated from water than are the terrestrial amphibia and to a large extent their success is due to their ability to restrict water loss from the body. The thick skin prevents water loss so effectively that animals such as iguanas and the rattlesnake lose water by evaporation only about 1/40th as fast as the fossorial toad *Scaphiopus*. Reptiles, like the amphibia, are poikilothermic and their metabolic and respiratory rate small. The low level of respiration aids in keeping down water loss in the respired air. Metabolic rate is affected by both temperature and activity however and hence exercise or a rise in body temperature result in increased water loss in the respired air. Thus the evaporative loss from an iguana goes up by some 400 per cent when the animal's environmental temperature is raised by 8° C. Exercise may raise the water loss by a similar amount in the lizard *Uma*. Most of this loss can be accounted for by the simultaneous six-fold rise in the respiratory rate.

Reptile renal tubules, particularly those of snakes and lizards, have rather small glomeruli. In a desert lizard such as *Trachysaurus* the total volume of the glomeruli (number times average volume) is only about 1/7th that of a comparable mammal, and hence the rate of urine formation is restricted. Indeed the kidney of the lizard is so adapted to reducing water loss that if the animal is over hydrated it takes about three times as long as a mammal to remove the excess water. As in other reptiles the kidney of desert forms cannot produce urine more concentrated than the blood. Hence if the water intake falls below the output or an appreciable amount of salt is ingested the animal's blood concentration rises. In the dry season water intake must frequently fall below output and the lizard *Trachysaurus* compensates for this by being very tolerant of variations in the blood. These animals survive a rise in their blood sodium of over 50 per cent, an increase far beyond that which could be tolerated by most vertebrates.

The solution to the water balance problems of terrestrial reptiles is thus based on restriction of water turnover and tolerance of raised blood concentration. The success in restricting water loss is clearly shown when a comparison is made between the rates of water turnover in lizards and rats and of comparable size on low water intakes. (Table 8).

TABLE 8. Comparison of the water losses from similar sized rats and lizards on low water intakes.

	Urine volume ccs/kg/day	*Evaporative loss ccs/kg/day*	*Total loss ccs/kg/day*
Lizard	less than 1	19	19
Rat	5	58	63

Data from Bentley and Dicker & Nunn.

The extreme reduction in urine volume in the terrestrial reptiles is only possible because the main waste product of nitrogen is uric acid. As it is so insoluble this substance can be eliminated as a sludge without raising the osmotic concentration of the urine. Reptiles which have a high water intake (and hence more water available for the secretion of smaller more soluble molecules) excrete either urea or ammonia. Thus though snakes, like the lizards, produce uric acid, turtles and tortoises are divided, the more terrestrial producing mainly uric acid whilst those from moister environments produce urea. The aquatic alligator produces a mixture of ammonia and urea.

Marine reptiles. The turtles, snakes, crocodiles and lizards all include members which inhabit or pay periodic visits to the sea. These forms, though hyposmotic to the medium are of course protected from excessive osmotic withdrawal of water by the impermeable skin. Nevertheless, they must have some means of disposing of salts taken in with the food and, as noted above, the reptile kidney cannot produce urine hyperosmotic to the blood. In the turtles, and lizard, excess salt is removed in 'tears'. It has long been known that marine reptiles 'cry' when they come ashore and there has been much speculation as to the cause.

One idea was that the tears served to wash the eyes free of sand, an unfortunate suggestion as the secretion merely makes the surface sticky and causes wind blown sand to adhere! The investigations of Schmidt-Nielsen have shown that these tears contain a higher concentration of salt than the blood. When the terrapin, *Malacolemys* is injected with a 10 per cent solution of NaCl it secretes tears with a concentration of 616–784 mE/l Na. Since the water in which it lives is very much less concentrated than this it is apparent that the animal could if necessary drink its medium and gain water by secreting the excess salt. This particular terrapin usually lives in brackish water and the fully marine Loggerhead turtle can secrete even more concentrated solutions. The tears of this animal have a concentration of almost 1,000 mE/l Cl and 900 mE/l Na. The marine iguana can do almost as well, as its tears contain up to 840 mE/l Na. These animals can therefore extract water from sea water (470 mE/l Na, 550 mE/l Cl) by eliminating the excess salts.

The orbital gland responsible for tear secretion consists in the Loggerhead and Iguana of closely packed tubules and is rather similar in design to that of the marine birds (p. 76). In other marine reptiles the glands are not so complex and the tubule arrangement less regular.

Mammals

Water is lost from the body of mammals by the same routes as in other terrestrial vertebrates namely by evaporation across the skin and in the respired air, urine and faeces. The skin of mammals is somewhat less heavily keratinised than that of many reptiles but insensible water loss is nevertheless almost as low as in the latter group. Despite this the overall evaporation from mammals such as rats is about 10 times that from reptiles of comparable size (Fig. 16). Two factors are involved in raising the loss from mammals both of which affect the removal of water in respired air. Associated with their homoiothermy* mammals have a high metabolic rate. A high metabolic rate necessitates

* Constant body temperature.

a rapid rate of breathing and, as we have seen in reptiles, this results in high loss of water in the expired air. The amount of water lost by a mammal is further increased because its temperature is usually above that of the environment. If the air which a reptile inhales is already saturated it will not lose water when it exhales. A mammal, even if it takes in air saturated at the ambient temperature, loses water in the exhaled air as this is saturated at the higher body temperature. This fact is admirably demonstrated on cold moist days as on each exhalation the excess water added by the lungs condenses out as the expired air is cooled.

Metabolic rate is related to body size, or rather to a function of surface area and volume, the smaller an animal the higher its metabolic rate per unit mass. Hence small mammals are particularly prone to high respiratory losses. For instance, when a man and rat are both short of water, the proportion of the daily water loss in the form of evaporation is about 70 per cent in the case of the larger and 90 per cent in the smaller animal. Activity increases the evaporation water loss since it raises the metabolic rate. Respiratory water loss declines as the saturation deficit of the air breathed decreases and we may suggest that this is one of the factors which explains why so many of the smaller mammals spend much of their time in damp burrows.

High metabolic rate necessitates a greater food intake which in turn requires a larger urine output to dispose of waste. Thus in addition to living their whole life at a higher tempo than large animals, small forms also have a more rapid rate of water turnover.

Mammals suffer from heat stroke when the body temperature is raised only 4 or 5 degrees centigrade above the normal level. Any tendency for the body temperature to rise is countered by the secretion of sweat or saliva on to the body surface. Evaporation of this water cools the body but the loss of water involved in this cooling may greatly aggravate the problem of maintaining the body water economy. As a considerable proportion of the water loss from the body is replaced in most mammals by drinking or eating succulent food, we might perhaps expect that the high temperatures and arid conditions of deserts would have prevented colonisation by members of this group. On the contrary, the capacity of a desert mammal, the camel, to do without water is legendary. In this section we will therefore consider the problems involved in the maintenance of water balance in a physiologically unspecialised mammal, man, and then go on to consider how the generalised mechanisms are modified in animals living in environments such as the sea and deserts where water is physiologically or actually in short supply.

Man. On a normal water and food intake the balance sheet for water uptake and loss in man has the following features. (Table 9).

TABLE 9.

	Losses		*Uptake*
Respired air	400 – 500 ccs/day	Metabolic water from general diet of 3,000 Cals	300–400 ccs
Insensible loss	500 – 800 ccs/day		
Urine	1,000 – 1,500 ccs/day	Water in food	1,000–1,500 ccs
Faeces	90 – 200 ccs/day	Drinking	as necessary
Sweat	0 – 12,000 ccs/day		
Total about	2,500 ccs/day	about	2,500 ccs/day

Naturally the losses in the respired air and insensible perspiration vary with the saturation deficit and activity, while the loss of water as sweat varies with the body temperature. These differences are discounted by varying the amount of water drunk. If water is not available for any reason the water output is decreased but the minimum daily loss is about 1,400 ccs. Of this compulsory loss some 1,000 ccs is lost by evaporation and the remainder is in the urine and faeces.

If a man is to remain in water balance he must, therefore, have an intake of at least 1,400 ccs per day. Some of this water can be obtained from the food. Fat, carbohydrate and protein all give some water on complete oxidation.

100 grams fat gives 105 grams water
100 grams carbohydrate gives 55 grams water
100 grams protein gives 45 grams water.

However, metabolic water cannot alone supply the water requirements of man however much he may eat as the processes involved in metabolising foodstuffs result in a greater loss of water than that gained. Man loses about 0·8 ccs of water in his expired air for each litre of oxygen consumed, and consequently the 200 litres of oxygen required to complete the metabolism of 100 grams of fat to CO_2 and water results in a loss of about 160 ccs of water. The net loss involved in metabolising 100 grams fat is thus about $160 - 105 = 55$ ccs. The oxidation of carbohydrate is somewhat less expensive of water to man but still involves a net loss of about 12 ccs of water per 100 grams substrate. Compared to the water loss on metabolising protein these losses are however trivial. When protein is broken down in mammals the main nitrogenous end product is urea and its removal therefore involves urinary water loss. About 30 grams of urea are formed for each 100 grams of protein metabolised. The maximum concentration of urea which can be produced in human urine is only 4·5 per cent so that the 45 ccs of water formed on protein oxidation is very small in comparison with the loss of water associated with the removal of the resultant urea.

A man must therefore supplement his solid intake with liquid water. This water must have a concentration lower than the maximum to which the urine can be concentrated if free water is to be available to make good the difference between water needed and that provided as metabolic water. The maximum concentration of the urine is 1·4 osmoles ($\triangle$ 2·6° C.) a level considerably above that of sea water ($\triangle$ 1·8° C.); but, as is well known, a man cannot survive if he drinks only sea water. The cause is multiple, partly due to the gut water loss caused by the irritating effect of the high concentration of magnesium in sea water, partly to the extremely slow elimination of excess salts and consequent over-hydration and over-concentration of the extracellular spaces, and partly to the fact that the concentration of chloride in sea water is more than the maximum concentration that can be excreted in the urine. The fact that the total concentration of urine may exceed that of sea water is obviously immaterial if the major anion of the water cannot be eliminated from the body as fast as it enters.

Progressive dehydration follows lack of drinking water; but the blood does not at once become markedly more concentrated, nor does its volume initially decrease. The maintenance of concentration and circulatory volume is achieved (1) by withdrawal of fluid from the interstitial spaces into the vascular system and (2) by a decrease in the volume and increase in the concentration of the urine. Loss of fluid from the blood concentrates the blood proteins and decreases blood pressure. These factors combine to ensure an osmotic withdrawal of fluid from the interstitial spaces (see also p. 154). The urine volume falls to 30 ccs an hour or slightly less and the salts and urea are maximally concentrated. Any further decrease in urine volume after the point when maximum concentration is reached would defeat its own end by decreasing the amount of salts and urea eliminated and hence raising the blood concentration. If net water loss is prolonged, elimination of water is no longer matched by salt removal and the concentration of the extracellular fluids begins to rise. This rise results in a passive withdrawal of water from the body cells until the osmotic concentration of cells and extracellular fluids is the same. Death occurs from the combined effects of increase in cell concentration and circulatory failure when about 20 per cent of the body water has been lost or when the blood concentration has risen from its normal level of 156 mM/1 NaCl to about 200 mM/l NaCl.

When water is not available there is a voluntary restriction of food intake in man and other animals. This self-imposed starvation may aid in the conservation of water since it results in a drop in metabolic rate and in faecal water loss.

High external temperatures or an increase in body temperature stimulate the secretion of sweat. Water loss in this form may reach very considerable proportions as up to 2 litres of sweat an hour may be produced over short periods. In man the sweat has a concentration of only about 1/10th to 1/3rd that of the blood so a rapid rate of sweating does not concentrate the body fluids so rapidly as, say, the panting of a dog, which results in a pure water loss. However, the fact that sweat contains salts may mean that there is a considerable drain on the body salt reserves if sweating is prolonged. If, after a period of heavy sweating, water alone is drunk to replace the fluid volume lost, the body fluids are diluted and muscular pains may develop (miner's cramp). For this reason miners, and others exposed to considerable losses of sweat, drink salted water (made acceptable in the form of beer).

A resting man fed on a dry diet would survive about 2 days at an air temperature of 38° C., but a camel has been found to live at least 17 days in such circumstances. How then do desert animals modify their water metabolism in order to survive in arid conditions? The means adopted depend on the size of animal and two species, the kangaroo rat and the camel, are chosen as examples.

The kangaroo rat. The Schmidt-Nielsens have found that the secret of the kangaroo rat's survival in a waterless desert is that it has contrived to restrict its water loss to an amount less than it obtains from the metabolic water of its food. Drinking is therefore unnecessary. Water loss is reduced by:

(1) Absence of sweating and tolerance of raised temperature.
(2) Production of a very concentrated urine.
(3) Production of dry faeces.
(4) Restriction of skin and respiratory losses.

These factors reduced the water loss associated with the intake of a given number of calories to a much lower level than that of a white rat of comparable size.

The absence of sweat glands is a common feature of small mammals whether or not they live in deserts. This may readily be explained when it is considered that in a very small animal any appreciable water loss will cause rapid concentration of the body fluids. Such animals therefore have the choice of dying of heat stroke or of dehydration if the ambient temperature rises above that of the body fluids. Consequently most small mammals avoid hot conditions by retiring underground during the heat of the day. To this rule the nocturnal kangaroo rat is no exception. As a compensation for the lack of sweat glands the kangaroo rat tolerates raised body temperature better than most mammals. It withstands temperatures up to 41° C., 6 degrees above its normal temperature. Normally, the burrows in which the animal lives are not only much cooler than this critical temperature, usually not more than about 30° C., but are also humid. This humidity aids in conserving water by decreasing the loss in the respired air. The important part played by humidity in the animal's water relations is clearly indicated in Fig. 12, which shows the losses and uptake of water of kangaroo rats fed on a diet of barley providing 100 cals (at different relative humidities).

In this figure the water losses in the urine and faeces are assumed to be constant at the minimum levels despite variations in the intake of water. The variations in water intake are due solely to the fact that the food is hygroscopic and hence its contained water varies with the humidity. The evaporative loss from the rat is seen to vary markedly with the humidity. In dry air evaporative loss accounts for over 70 per cent of the total loss, at 80 per cent relative humidity it is only about 40 per cent. At 10 per cent relative humidity the line demarcating total loss cuts the line which indicates total intake. At all humidities above 10 per cent therefore the animal can keep itself in water balance merely by making the appropriate increase in urine output. At relative humidities above 20 per cent, intake from the metabolic water alone balances the losses and the animal can then live on perfectly dry food. In contrast a laboratory white rat can only balance uptake and loss on a similar diet without drinking if the relative humidity is above 95 per cent.

Part of the kangaroo rat's success in reducing evaporative loss is thought to be due to the evaporation of a small quantity of water from the nose. This cools the nasal passages and hence the expired air is saturated with water at a temperature lower than that of the body.

The faeces are very dry containing only about one third the amount of water per dry weight found in the faeces of a white rat.

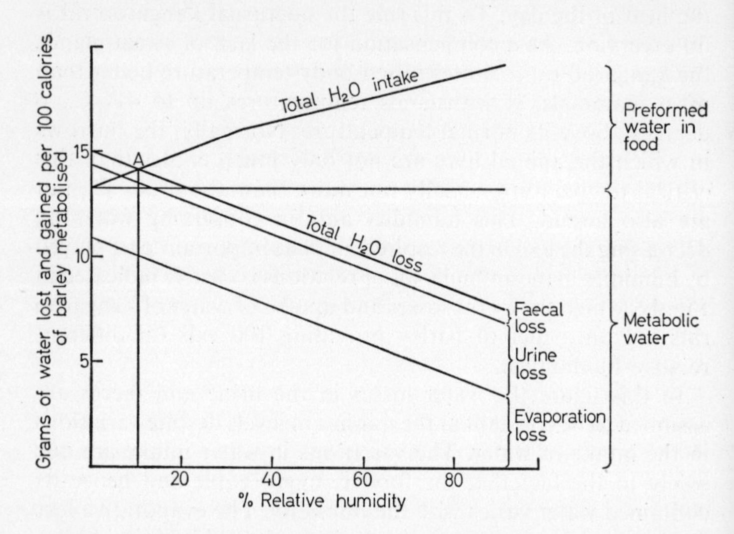

FIG. 12. The effect of the relative humidity of the air on the water relations of the kangaroo rat, *Dipodomys*. For the purpose of constructing the diagram it is assumed that the urine and faecal water losses remain constant at the minimal levels whatever the humidity. The point 'A' indicates the humidity at which water intake and loss are equal. (After Schmidt-Nielsen, K. and Schmidt-Nielsen, B.)

The urine is highly concentrated. A level of up to 5·6 M/1 NaCl has been recorded, four times as concentrated as the maximum of man and about five times the concentration of sea water. When it is fed on a high protein diet large amounts of urea are formed and eliminated and the kangaroo rat is no longer able to restrict its water loss to the level obtained as metabolic water;

it must then drink. However, the kidney can concentrate the urine to such an extent that even sea water can be drunk to make up the deficit.

The powers of water conservation are so well developed in the kangaroo rat that it does not normally suffer wide changes in blood concentration or body water content and no unusual tolerance of these effects is present. The animal dies when the water content is reduced to about the same level as that required to kill other mammals.

The camel. Large animals living in deserts and steppes are obviously unable to emulate the kangaroo rat and take cover in humid burrows. Hence they are exposed both to the heating effect of the sun and to relatively dry air.

In assessing the ability of these animals to survive in such conditions we must first dispose of any suggestion that they have special water stores in either the stomach or hump. Fat, it is true, is present in considerable amounts in the camel's hump and metabolic water is derived from its oxidation. However, the camel is unable to respire air which is initially at a high humidity and we have no grounds to suppose that it can restrict water loss in the expired air. It is therefore improbable that the camel, any more than man, can derive a net surplus of water from fat metabolism. The fat store is therefore best regarded as a store of metabolic energy which will sustain the animal over the long distances which must be travelled between sources of drinking water. Another suggestion has been that water is kept in pouches in the stomach and released as required, but recent reports deny that water in the stomach can be utilised as a reserve in this way.

When short of water a camel decreases its daily urine volume to about 5 litres, a small but not exceptionally low level. The maximum concentration too is not particularly high, only a little over half that of the kangaroo rat.

Sweat glands are present in the camel. In winter when the vegetation is fairly succulent, a camel has been known to go 60 days without drinking and even in summer a resting animal in full sun survives at least 17 days on dry food. The daily percentage water turnover of a camel in winter is not dissimilar to that of a man. Comparison is made in Fig. 13 of a camel, donkey, man and kangaroo rat, showing that only the last has

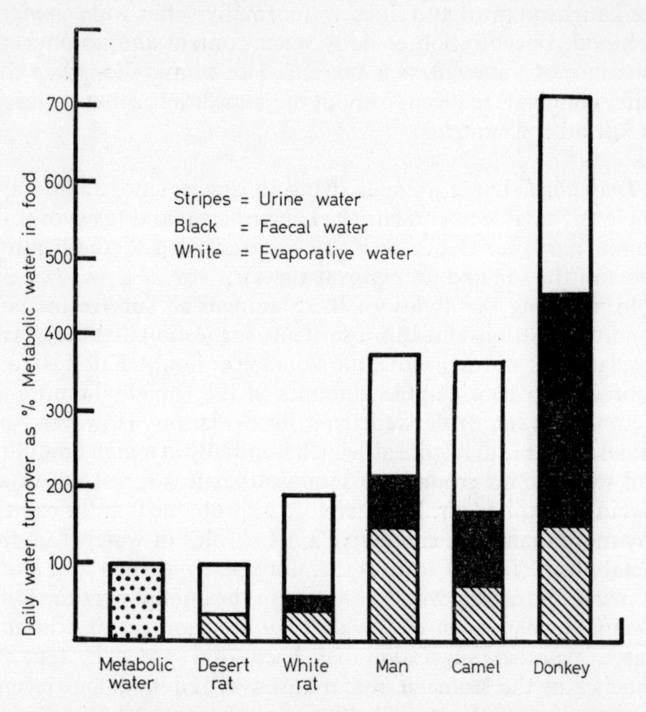

FIG. 13. Comparison of the daily water turnover of various mammals to show that only the desert-living Kangaroo rat, *Dipodomys* can keep in water balance on metabolic water alone. (After Schmidt-Nielsen *et al.*)

a water turnover within the amount provided by metabolic water.

If the camel does not store water, sweats, has a high respiratory loss and does not produce a particularly concentrated urine you might ask how it manages to survive so long between drinks. Schmidt-Nielsen and his colleagues have again produced most of the answers. The two features in which a camel differs from most other mammals are (1) its tolerance of changes in body temperature and (2) its tolerance of change in body water content. When the animal is fully hydrated and has an adequate water intake it sweats to control its body temperature. When the body water content falls, sweating is suppressed and a rise of body temperature of up to 6·2° C. is tolerated. The water loss which would have been required to dissipate this stored heat if the body temperature had instead been kept constant is thus saved. For a 400 kilogram camel the saving of water is considerable, of the order of five litres a day. During the night the stored heat is radiated from the body and the temperature falls back to the normal level. In the early morning vascular changes occur which increase the rate of heat loss and so depress the body temperature below normal in preparation for the rise during the day.

At first sight it might appear curious that desert animals such as the camel and donkey should have thick coats of fur. However this fur acts as an insulator slowing down the rise in body temperature during the day. When a camel is shaved it loses substantially more water than one which has its coat intact.

Perhaps the most important single feature of the camel's water metabolism is its considerable tolerance of water loss. It will survive a water loss which lowers its body weight by 30 per cent. This is ten per cent more than the loss which would prove fatal to man and most other mammals.

Man when allowed access to water after a period of dehydration does not at once drink sufficient to account for his loss. The camel on the other hand takes in just the right amount of water to balance its loss and furthermore takes it in with great rapidity. One camel after a loss of 103 kilograms (181 pints) took in just this amount of water in 10 minutes, a capacity that some topers might envy!

When a camel is working in the heat of the day the increased metabolic rate aggravates the tendency for the body temperature to rise and increased evaporation becomes necessary. A working camel cannot survive on dry food alone and to be kept in good condition must be watered about every three days.

A donkey tolerates about as great a water loss as a camel, but, it loses water some two and a half times as rapidly as the latter. The camel can therefore survive about two and a half times as long as a donkey in comparable conditions.

Summarising differences in water balance regulation in the kangaroo rat and the camel we may say that the former requires only the metabolic water in its food, the latter must have succulent food or drink at intervals. In the rat respiratory loss is reduced by cooling the expired air and by the fact that it spends much of its time in burrows where the humidity is high. The urine is small in quantity and exceptionally concentrated. The camel cannot avoid the heat of the day and its survival depends largely on tolerance both of high body temperatures and the loss of water. Fig. 13 compares the routes of water loss in man, camel, donkey, white rat and desert rat and indicates that only the last can keep in water balance on its metabolic water.

Marine mammals. Marine mammals, like marine reptiles, maintain a blood concentration similar to but a little higher than that of the terrestrial members of their group. (Table 10). They

TABLE 10.

Blood concentration

	△ °C		△ °C
Sperm whale	·66 – ·72	Sheep	·59
Phoca vitulina (seal)	·54 – ·61	Dog	·56
Dolphin	·83	Man	·57

are thus very hypotonic to their medium. As in the reptiles, the impermeable skin of mammals prevents any gross withdrawal of water across the body surface. Sweat glands are absent (sweating would be useless in a submerged animal as it is of course the process of evaporation which cools the body and not the secretion of the fluid). Respiratory loss too is likely to be fairly low as the air breathed, being taken from close to the surface of the sea, will have a fairly high humidity. Evaporative losses are relatively low, therefore, leaving the main sources of loss as the urine and faeces. In a seal, evaporative losses are under 10 per cent of the total water loss, a complete reversal of the situation in a white rat, where evaporative loss can be 90 per cent of the total.

If marine mammals can therefore produce urine with a concentration of ions only 10–20 per cent more concentrated than in their fluid intake and can at the same time dispose of the nitrogenous waste of protein intake, they will be in water balance. The extent to which the urine must be concentrated to achieve this end depends on the nature of the food. Seals, dolphins, porpoises and the large toothed whales are mainly fish eaters though the toothed whales certainly supplement their diet with squid. The whale-bone whales subsist on invertebrates, mainly crustacea 'filtered' from the plankton. The blood of teleosts is hyposmotic to sea water but still hyperosmotic to the blood of mammals whilst the invertebrates are probably all isosmotic with sea water. None of the mammals can gain water therefore without performing osmotic work; but the amount of work depends on whether the food is fish or invertebrate. It is important therefore to know if the kidney of the mammal can produce a sufficiently concentrated urine to dispose of waste from the food or whether its action must be supplemented by an additional salt secreting organ as in the reptiles.

Some values for the concentrations of urine in different forms are shown in Table 11.

TABLE 11. The Composition of the Urine of Marine Mammals

	△ °C	*urea mM/l*	*Cl mE/*1
Sei whale	2·49	400	370
Blue whale	2·50	420	340
Porpoise	2·3–3·4	480–840	170–240
Seal	1·7–4·0	120–1,050	200–420

Even the highest of these values are not strikingly different from the maximum concentrations that can be achieved by man △° C. 2·6, urea 700 mM/l, Cl 370 mM/1. As the blood chloride concentration of invertebrates is about 500 mE/1 it might appear at first sight that these levels are inadequate even to remove the ions taken in. It must be remembered, however, that the total inorganic ion concentration of the tissues of the food organisms is lower than that of the blood (p. 87) and that chloride in particular is present in very much lower concentrations than in the blood. Even if a fairly generous allowance is made for the extracellular volume the chloride concentration taken in the total body water of an invertebrate is unlikely to exceed about 375 mM/l. Even the fish-eating seals, which will usually have a much lower chloride intake, could dispose of this level of chloride and we may suppose therefore that this is well within the powers of the animals which normally feed on invertebrates. The urine chloride values given for whales are those observed when the animals were killed. The maximum capacity of various whales to concentrate ions in their urine may be above this but has yet to be determined. It would clearly be disadvantageous for the marine mammals to drink sea water as this would necessitate the removal of ions at a much higher concentration than that calculated above. Apparently they do not drink. Irving and his colleagues have carried out an experiment that shows convincingly that the concentration of urine which must be produced by seals is within the range of values observed. They kept seals for four months on a diet of herring. Knowing the amount of herring consumed they calculated the water uptake and loss for each 1,250 grams of fish eaten. This food contained about 1,120 ccs of water (metabolic and as liquid). Respiratory losses accounted for 106 ccs and faeces for a further 200 ccs which left rather over 800 ccs for urine formation. Some 50 grams of urea and 12·5 grams of salt were liberated from the food and if dissolved in the water available for urine production would give a concentration of about △ 2·7° C. This is well within the capacity of the seal.

We may conclude that there is no reason to suppose that the kidneys of marine mammals feeding on invertebrates need supplementing by the activity of any additional salt secreting glands.

Birds

Birds have a high metabolic rate and their temperature is usually maintained at a level somewhat above that of mammals. Consequently, water loss in the respired air is considerable and forms an important part of the total water loss. Larger birds such as owls lose about 5 per cent of their body weight per day via this route by comparison with the 1 per cent lost by man; but respiratory losses increase with decreasing body size and in a small bird such as a wren, may amount to as much as 35 per cent of the body weight per day. Even desert birds have developed no means of restricting this loss and lose more water in the respired air alone than they gain in the form of metabolic water from their food. Hence all birds must have access either to liquid water or to succulent food. In response to the need for a high water intake many of the smaller birds are insectivores while the smallest birds of all, the humming birds, feed primarily on liquid nectar. Humming birds have about twice the oxygen consumption of wrens and presumably have proportionally greater respiratory water losses. Even a diet of nectar would be insufficient to keep humming birds in water balance without other modification to normal metabolism. Since humming birds are so small the ratio of surface to volume is large and consequently heat is rapidly dissipated from the body. The maintenance of a high body temperature therefore necessitates an exceptionally large metabolic rate. Associated with the high metabolic rate oxygen consumption is considerable (about 85,000 ccs O_2/kg. body weight/hour in a flying Allen humming bird, in comparison with 200 ccs/kg/hr in a resting man). If humming birds were to maintain the body temperature at its normal level when that of the environment fell at night, not only would the call on metabolic reserves be considerable, but the animals would be appreciably dehydrated by morning. Consequently, many humming birds and particularly those inhabiting high mountain forests, become poikilothermic ('cold blooded') at night. In effect, they go into a state of temporary hibernation. In the case of the Anna humming bird the oxygen consumption falls during the small hours to a level only about one tenth of that during the day. The changes in water loss associated with this drop in respiratory rate have not been studied but may be expected to be proportional. A parallel can be drawn here with the bats, which are similarly exposed to high respiratory water losses and which also tend to lower the body temperature when inactive.

Birds compensate for their high respiratory water loss by decreasing the loss by other routes. Urine is not released directly to the outside but, as in reptiles, the ureters empty into a cloaca. The urine volume itself is small and may be still further reduced after leaving the ureters, since the wall of the cloaca is able to take up water from the urine and faeces. Doubtless there is some insensible loss but water loss through the skin is limited by the absence of sweat glands. Despite these restrictions the water turnover of birds is still considerable. Even as large a bird as a $5\frac{1}{2}$ lb hen exchanges about 14 per cent of its body water per day when on a normal diet, a comparable figure for man being about 3·5 per cent. In the case of the hen the total loss is approximately equally divided between the urine and faeces on the one hand and evaporative and respiratory losses on the other. Metabolic water accounts for about 30 per cent of the intake and the remaining 70 per cent has to be taken as liquid.

The renal tubules of birds have loops of Henle (see p. 119) and they are the only group of vertebrates other than the mammals which are able to form urine hyperosmotic to the blood. However, the capacity of the bird kidney to concentrate ions is not great and the osmotic concentration of the urine never approaches that found in the more specialised mammals. The maximum concentration of the urine of a hen, or of a marine bird such as a cormorant, is equivalent to 0·3M/l. NaCl, only twice the concentration of the blood, while the urine of the desert-living kangaroo rat may be 19 times as concentrated as its blood. The bird kidney-rectal system in fact is like that of the reptiles in being adapted to produce a very small volume of urine rather than a very highly concentrated solution.

Again like the reptiles this restriction of urine volume is made possible because the main end product of nitrogen metabolism is uric acid. By secreting this substance into the renal tubules birds can achieve the same proportion of nitrogen to water loss in the urine as the desert rodents despite the lower osmotic concentration and small fluid volume of their urine. Only sufficient fluid need be present to sweep the uric acid sludge into the cloaca.

The small urine volume and low concentration aids the retention of salts in the body. This feature may be of considerable importance to birds of inland regions, particularly seed eaters, but to marine birds is positively disadvantageous. Marine birds, unlike submerged marine animals, such as mammals and reptiles, have a high evaporative water loss. Many marine birds drink sea water to compensate for this loss but this means that the salts of sea water must be excreted at a level exceeding that in sea water itself, a capacity quite beyond the bird kidney. The task of eliminating excess salt is taken, as in the marine reptiles, by glands in the region of the orbit.

It has long been known that the bilateral nasal glands are very much larger in marine than in inland birds and it was formerly supposed that the glands secreted a fluid to protect delicate membranes of the nasal canal from the high concentration of salts in sea water. The discovery by Schmidt-Nielsen and his colleagues that the secretion of the nasal gland of the cormorant, penguin and herring gull has a higher salt concentration than sea water reveals at once the dubious nature of the previous hypothesis and the true function of these 'salt' glands. The glands are active only when the salt concentration or osmotic pressure of the body fluids is raised. In the cormorant they then secrete about one cc of fluid every five minutes with a concentration of 500–600 mE/l. Na, four times the concentration of sodium in the blood. If this rate of secretion were to be continued indefinitely, all the salt in the body fluids would be put out in about 10 hours. The Humbolt Penguin has an even more effective salt gland since it can produce up to 0·36 ccs a minute of a fluid with a concentration of 700–840 mE/l Na. When five grams of sodium chloride were fed to a penguin the bird eliminated two thirds of the salt in four hours in 80 ccs of nasal gland secretion. By comparison the kidney excreted only about one tenth of this amount of salt in the same period. So effective is this salt gland secretion that rather over one third of any sea water drunk by the penguin could be made available as pure water for other purposes.

The salt gland of the herring gull is composed of branching tubules radiating outwards from a central canal. Large arteries supply the gland but a shunt is available as a bypass when the gland is not secreting. (Fig. 14).

Similar glands have been found in marine birds belonging to a number of different families. These include gulls, ducks, pelicans, petrels and penguins. In the gulls there is a correlation between the size of the gland and the extent to which the species is associated with marine conditions. The herring gull, which is practically confined to the littoral region, has a large gland. The common gull frequently penetrates up estuaries and inland and has a rather smaller gland, whilst the black-headed gull, a frequent coloniser of inland waters, has a comparatively small gland.

SUMMARY OF VERTEBRATE REGULATION

Fresh water forms

These fall into two groups, the fish and the amphibia, part of whose body surface is permeable to water, and the reptiles and mammals, whose surface is effectively impermeable. The fish and amphibians take up water by osmosis and secrete a copious and dilute urine. Ion loss is replaced by active uptake through the gills in fish and over the whole body surface in amphibia. The reptiles and mammals have no major osmoregulatory problem as water uptake and ion loss through the skin is small. Ion uptake through the skin does not occur and all salts are taken in with the food.

Marine forms

The marine teleosts are hypotonic to the medium. They lose water by osmosis and take up ions by diffusion. Sea water is swallowed, NaCl and water absorbed from the gut and excess salt eliminated by the gills. Mg, Ca and SO_4 pass out either in the faeces or urine.

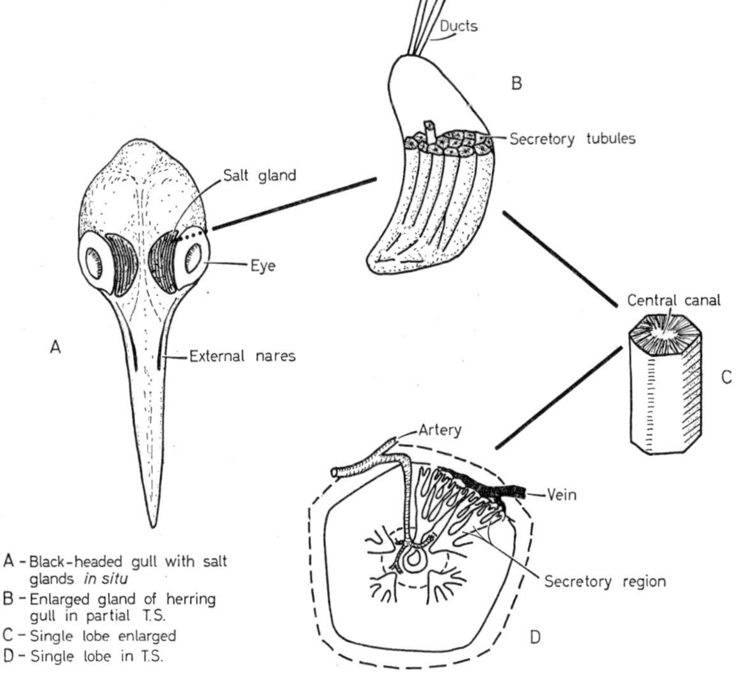

FIG. 14. The salt-secreting glands of gulls. (Mainly after Fänge, Schmidt-Nielsen and Osaki)

Elasmobranchs have a salt concentration little higher than that of teleosts but retain so much urea that they are slightly hyperosmotic to sea water. Excess NaCl diffusing in or taken in food is removed by the rectal gland.

Amphibians are generally intolerant of high salinities. The only known species to live in high salt concentrations retains urea in a manner similar to the elasmobranchs.

Marine reptiles are hypotonic to sea water but the impermeable skin limits osmotic water loss. The kidney cannot produce hyperosmotic urine and hence is not able to remove excess salts taken in with the food. Sea water is drunk to replace fluid lost in the urine, faeces and respired air. Excess salt is secreted in highly concentrated fluid from orbital glands.

Mammals, like the reptiles, have low evaporative losses and suffer little from osmotic water loss. Unlike the latter group they produce urine hyperosmotic to the blood. This is adequate to remove excess salts from the body. Mammals do not normally drink the medium as this would increase the amount of salt to be removed.

Birds are the only marine vertebrates to suffer high evaporative losses. The kidneys can secrete urine hyperosmotic to the blood but this is still inadequate to eliminate the salts from any sea water drunk. Excess salt is removed via nasal glands.

Terrestrial forms

The relation between evaporative losses and water loss by other routes depends on the permeability of the body surface, size, body temperature, the nature of the main nitrogenous end product and the saturation deficit of the surroundings.

Amphibia have a relatively high surface permeability and must remain in regions or local niches where the humidity is high. Nitrogenous waste is removed mainly as urea, and during temporary water shortage, urea can be retained in the body. The kidneys cannot produce urine hyperosmotic to the blood.

Reptiles reduce surface evaporation by having an impermeable skin. The kidney cannot produce urine hyperosmotic to the blood but nitrogenous wastes can be removed in a very small volume of urine as uric acid is the main end product of fully terrestrial forms. Respiratory water loss is related to metabolic rate and hence to activity and temperature.

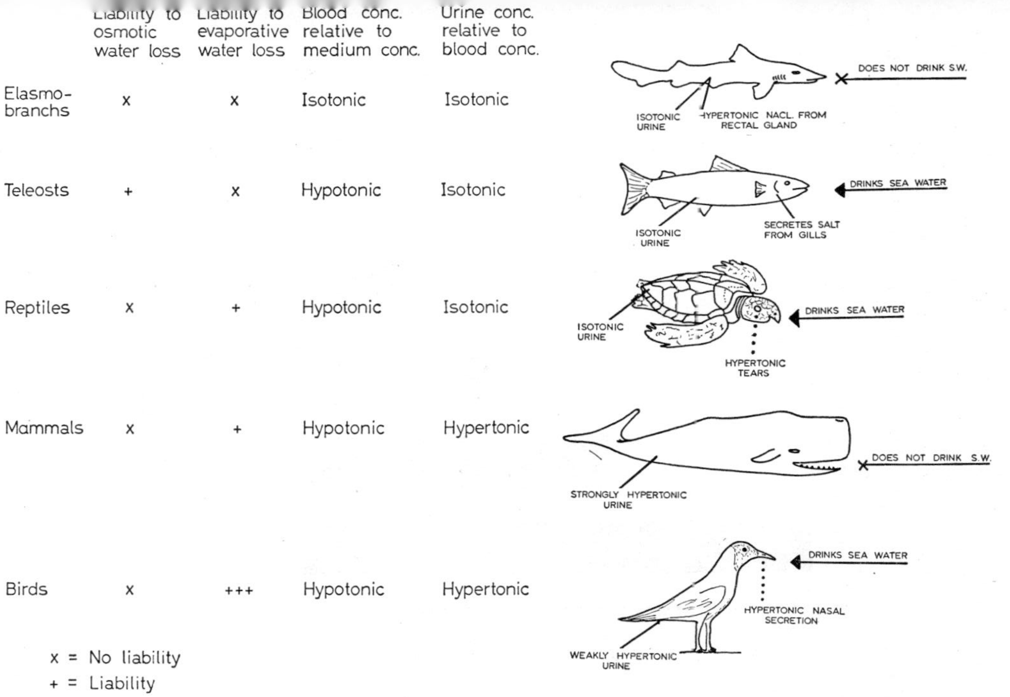

	Liability to osmotic water loss	Liability to evaporative water loss	Blood conc. relative to medium conc.	Urine conc. relative to blood conc.
Elasmo-branchs	x	x	Isotonic	Isotonic
Teleosts	+	x	Hypotonic	Isotonic
Reptiles	x	+	Hypotonic	Isotonic
Mammals	x	+	Hypotonic	Hypertonic
Birds	x	+++	Hypotonic	Hypertonic

x = No liability
\+ = Liability

FIG. 15. The various methods used by marine vertebrates in the regulation of their salt and water balance.

Mammals also have a skin with a low permeability to water but the high body temperature makes the evaporative losses from the respiratory passages considerable. Urea is the main end product of nitrogen metabolism. The urine can be made

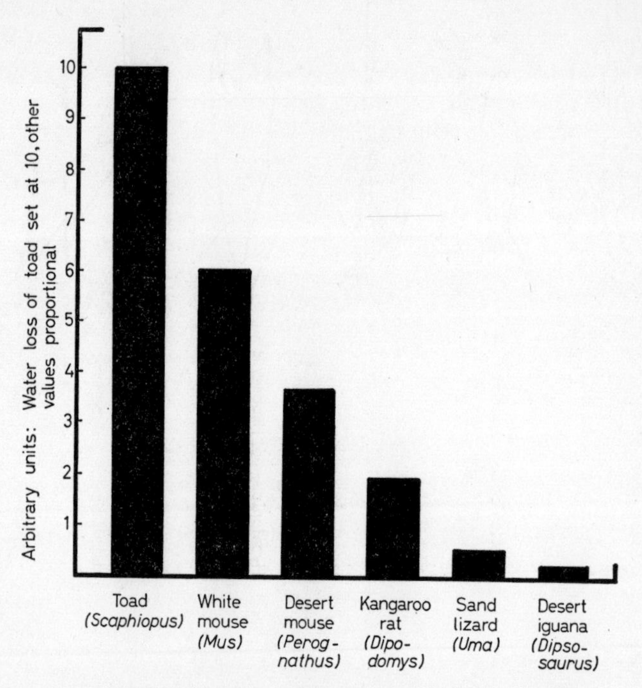

FIG. 16. Comparison of the rate of water loss from various terrestrial vertebrates of similar size at the same ambient temperature and humidity. (Data for diagram from Chew and Dammann)

hyperosmotic to the blood. Water loss is increased by activity and if the body temperature rises.

Birds have a high metabolic rate and their respiratory losses form a large part of the total water loss. The urine is hyperosmotic to the blood and urine volume need not be large as the main nitrogenous waste is uric acid.

In Fig. 16 a comparison is made of the evaporative losses in various similarly sized terrestrial forms. This figure shows (a) the effect of high skin permeability (comparison between toads and reptiles; (b) the effect of body temperature above that of the environment (comparison between reptiles and mammals), and (c) the considerable reduction in evaporative losses shown by desert mammals in comparison with the white mouse.

4

The Ionic Composition of Blood

MARINE FORMS

With the exception of a few prawns, crabs, and isopods the blood of marine invertebrates is isosmotic with sea water. The hagfish *Myxine* is also isosmotic. The main part of the blood osmotic pressure is contributed by inorganic ions. The most important of these, as in sea water, are sodium and chloride. Most of the other ion species in extracellular fluid are also present in proportions which at a first approximation resemble the levels in sea water (Table 12), a factor which is perhaps hardly surprising when it is considered that the earliest living cells must have been adapted to this medium and that the composition of sea water is not now thought to have changed grossly, at least

TABLE 12. Ion concentrations of the body fluids of three marine species as per cent of the levels in sea water.

		Na	*K*	*Ca*	*Mg*	*Cl*	SO_4
Echinoderm							
Marthasterias glacialis	coelomic fluid	100·1	111·2	101·0	97·7	100·8	99·7
Annelid							
Arenicola marina	plasma	100·1	103·5	99·8	100·3	99·7	92·2
Arthropod							
Carcinus maenas	plasma	111·0	121	127	35·8	99·8	57·2

From Webb and Robertson.

over the last 500 million years. Nevertheless, no animals have blood whose ionic composition is identical with sea water and in most cases the more highly organised and active animals show the greatest distinction between their blood and the medium.

Echinoderms, which have both a relatively low level of organisation and circulate the medium through their water vascular system, have coelomic fluids which differ only in minor details from sea water (though even in these animals there is a tendency to concentrate potassium within the lantern coelom). More active forms such as the crustacea tend to increase the potassium level in the blood and to reduce the levels of magnesium and sulphate. Even within a single group of crustacea, the decapods, there is a good correlation between the magnesium levels in the blood and the general activity of the species. Animals such as the spider crab *Maia* which are rather sluggish have a high level of blood magnesium, less Mg is present in the more active *Carcinus*. It is possible that there is a direct correlation between the concentration of magnesium and activity as it is known that this ion can interfere with the transmission of impulses at the neuromuscular junction. Care must be taken however when making such generalisations. In this case it is possible that the effect may be applicable to the crustacea. It is not applicable to the equally active cephalopod molluscs all of which have blood magnesium levels not dissimilar to those in sea water.

Three factors are involved in maintaining differences between blood and medium ionic concentration: (1) the continuous formation of urine; (2) the Donnan equilibrium, and (3) the active transport of ions.

The influence of urine formation

Many years ago it was pointed out that even if an animal produces urine which has an ionic composition like that of the blood, then any ions which diffuse across the body wall more slowly than they are removed in the urine will have their concentration in the blood depressed. Magnesium, calcium and sulphate diffuse more slowly across biological membranes than smaller ions such as sodium, potassium and chloride and hence their concentration in the blood might be lowered as a result of urine formation. In fact the excretory organ does not play a purely passive role even in those forms which produce urine isosmotic with the blood. Ions which are present in lower concentrations in the blood than in sea water are usually more concentrated in the urine than the blood. Modification of the urine volume too may result in changes being made in the ionic composition of the urine. When *Carcinus* is kept out of water the urine volume is reduced and at the same time the magnesium concentration is increased and the sodium concentration is decreased. (Fig. 8).

Some means must be present to ensure that the fluid and smaller ions lost in the urine are continuously replaced if urine formation is to play a part in regulating the concentration of the divalent ions. Furthermore, as diffusion cannot concentrate the smaller ions in the blood relative to the levels in sea water some other mechanism must account for the raised concentration of these ions in the blood.

The Donnan equilibrium

The blood of all animals contains protein in solution and this would be expected to influence the distribution of ions across a permeable body wall. When a large charged particle (such as a protein) is separated from an isosmotic solution of an electrolyte by a membrane through which the ions of the electrolyte, but not those of the large particle, can penetrate, then a redistribution of ions occurs. Thus, if the two solutions initially separated by such a membrane were a sodium proteinate and a mixture of sodium chloride and potassium chloride, the condition at ionic equilibrium would be such that $\frac{Na_i}{Na_o} = \frac{Cl_o}{Cl_i} = \frac{K_i}{K_o}$ where the suffix i indicates the concentration of ion on the same side of the membrane as the protein and o that on the other side. The presence of the protein tends to concentrate the inorganic cations relative to the anion on the protein side of the membrane. However, if the blood concentration of marine animals were maintained different from that of sea water by the establishment of such Donnan equilibria then the dialysis of blood against sea water through an ion permeable membrane should not affect its composition. In fact there are marked changes in the composition of the blood of several crustacea when the blood is thus dialysed (Table 13). The normal composition of the blood must therefore be at least in part due to the active transport of ions against their activity gradients.

TABLE 13. The concentration of ions in the plasma of some marine crustacea expressed as a percentage of the concentration in plasma which has been dialysed against sea water.

Species	*Na*	*K*	*Ca*	*Mg*	*Cl*	SO_4
Eupagurus bernhardus	105	130	137	49	96	135
Maia squinado	100	125	122	81	102	66
Portunus puber	110	147	120	41	101	83

Data from Robertson.

Active transport

The active transport of ions necessitates the expenditure of energy. Theoretically the work done depends on the amount of ion transported and the concentration gradient through which it is moved. The minimum work which must be done to transport 1 gram mol of an ion from a concentration x to a higher concentration y is W cals where $W = R.T.\ 2{\cdot}303 \log_{10} \frac{y}{x}$. R is the gas constant and T the absolute temperature. It is probable that the mechanisms responsible for active transport are not 100 per cent efficient so the actual work will be in excess of this.

NON-MARINE AND HYPOSMOTIC MARINE FORMS

Non-marine animals (with the exception of some insects) and hyposmotic marine forms maintain the same general ionic ratios in their body fluids as marine animals. As they are not isotonic with their medium the maintenance of the ion concentrations depends on regulation of the rate of ion uptake and loss. Active transport of ions in non-marine forms and the factors influencing the rates of transport are considered in Chapter 6.

SUMMARY

The composition of blood of marine invertebrates and *Myxine* is broadly similar to that of sea water. The main differences are a tendency to decrease magnesium and sulphate concentrations and increase the level of potassium. Three mechanisms are responsible for maintaining the differences between the blood and medium. (1) Urine formation; (2) the Donnan equilibrium; (3) the active transport of ions.

5

The Ionic Composition of Cells

It has long been known that the ash content of cells of isosmotic marine invertebrates is only about one third to one half that of the plasma and hence that a lower proportion of inorganic ions must be present in the tissues than in the blood. The inorganic content of the cells forms a larger portion of the total cell osmotic pressure however in forms which have lower blood concentrations. Thus the hagfish *Myxine*, the Roman eel, *Muraena*, and the rat form a series in which the organic contribution to the cell osmotic pressure becomes progressively less important. (Table 14).

TABLE 14.

		Muscle-cell O.P.	
	Blood O.P.	*% inorganic ions*	*% organic*
Myxine	100	42	58
Muraena	35·4	72	28
Rat	26·5	87	13

Myxine osmotic pressure set equal to 100, *Muraena* and rat osmotic pressures proportional.

Data based on values in Robertson.

The composition as well as the ionic concentration of cells differs from that of the extra-cellular fluids. Table 15 showed that the main ions of the blood are sodium and chloride while potassium, magnesium and phosphate are of lesser importance. The same is also true of the interstitial fluid. In contrast the main inorganic ions of cells are potassium, phosphate and magnesium. Comparison of the ionic content of the blood and cells of the Norway lobster *Nephrops* illustrates this difference. (Table 15).

TABLE 15. Comparison of the intra- and extra-cellular ionic composition of the Norway lobster, *Nephrops*.

	mg.ions / kg. water							
	Na	*K*	*Ca*	*Mg*	*Cl*	SO_4	HCO_3	*P*
Intra-cellular	24·5	188	3·72	20·3	53·1	1·02	8·8	164·2
Extra-cellular	517	8·6	16·2	10·4	527	18·7	4·13	0·81

Data from Robertson.

However not all inorganic substances in the intra-cellular fluid are freely ionised. Calcium and sodium are inhibitory when present in high concentrations inside cells and calculations made on *Nephrops* muscle indicate that a large proportion of these ions inside the cells are bound to other constituents.

Even within a single individual the intracellular ion concentrations vary with the type of cell examined. Thus slow muscles in molluscs have higher sodium and chloride concentrations and lower potassium concentration than fast muscles, (Table 16). The same is true of striated and smooth muscles in vertebrates.

TABLE 16. Comparison of the ion concentrations of slow and fast muscle in *Mytilus* to show that the ionic composition of cells is not uniform even in the same individual.

	mE/kg · fibre water		
	Na	*K*	*Cl*
Slow muscle	95	134	152
Fast muscle	79	152	94

Data from Potts.

At one time it was assumed that the differences in the ion concentration of the cells and blood could be maintained because the cell membranes were impermeable to inorganic ions. However, when radio-active isotopes became available it was soon shown that tracer labelled sodium or potassium placed in the extracellular fluid readily penetrated into cells. The cell wall cannot therefore be impermeable and a continuous exchange of ions takes place between blood and cell. Sodium tends to leak into cells down the concentration gradient and potassium would leak out if its movement were influenced only by its concentration gradient. The differences between ionic composition of cells and their bathing medium thus has to be actively maintained. The most important of the mechanisms involved is believed to be that responsible for extruding sodium from cells as fast as it diffuses in down the electro-chemical gradient. However, a charged ion such as sodium cannot be transported out of a cell unless an ion of similar charge is moved in or one of opposite charge is moved out at the same time. The anions of cells are composed principally of large organic molecules which cannot readily pass the cell membrane. Consequently, when sodium is extruded exchange occurs for potassium from the blood. Continuous extrusion of sodium results in the normal state whereby potassium is the principal cation associated with the organic anions of the cell. Now potassium would tend to diffuse out of the cell down its concentration gradient were it not prevented by the inability of the associated anions to escape. Since the anions cannot escape the cation, potassium, can only be lost until the potential built up by the separation of charge across the membrane offsets the tendency for it to diffuse down the concentration gradient. Normally, therefore, this effect creates a potential difference across cell membranes such that the inside of the cell is negative to the outside. In vertebrate muscles and nerves the magnitude of this *membrane potential* is approximately related to the ratio of the concentrations of potassium in the blood and cells:

$E = \frac{kT}{e} \ln \frac{K_i}{K_o}$ where E is the membrane potential, K_i is the potassium concentration of the extracellular fluid, T is the absolute temperature and k is a constant. At 20° C. $\frac{kT}{e}$ has a numerical value of approximately 25·1 and $E = 58 \log_{10} \frac{K_i}{K_o}$.

If the extracellular concentration of potassium is raised the membrane potential falls and if it is lowered the potential rises.

Chloride distribution between cell and extracellular fluid is controlled by the membrane potential. Chloride enters the cell until the concentration and electrical gradients along which the ions are moving are equal and opposite. Exchange of chloride between cell and extracellular fluid continues after this equilibrium point is reached but there is no further net movement. If the cell membrane were only permeable to potassium and chloride, the latter would come into electrochemical equilibrium when the conditions of the Donnan equilibrium were satisfied such that

$$\frac{K_i}{K_o} = \frac{Cl_o}{Cl_i}$$

Where K_i and Cl_i are the concentrations of potassium and chloride in the cell and K_o and Cl_o the concentrations in the extracellular fluid.

In many vertebrate muscles this condition is found approximately. The same is also true for many invertebrate muscles such as those of *Carcinus* leg muscles and *Mytilus* heart muscle. In some more slowly contracting mollusc muscles however and in the muscles of the Norway lobster *Nephrops* and the Chinese wool-handed crab *Eriocheir*, the Donnan ratios are not obeyed, $\frac{K_i}{K_o}$ being greater than $\frac{Cl_o}{Cl_i}$ and in these cases it is probable that potassium is actively taken up by the cell.

The relation between the size of the resting potential and the ratio of potassium concentration inside and outside cells is dependent on the fact that cell membranes are usually very much more permeable to potassium than to sodium. Any increase in the permeability to sodium tends to decrease the resting potential. This is because there is then a tendency for sodium to move down the gradient and offset the original resting potential. If the membrane were to be more permeable to sodium than potassium then a situation would eventually be reached when the sodium was tending towards electrochemical equilibrium. (Since the sodium is diffusing in the opposite direction to potassium the potential would then have the reverse of its normal polarity.) This effect is utilised in nerves and muscles to create the *action potential.* Stimulus of a muscle or nerve cell is followed by a rapid change in the permeability of the membrane to sodium. In the resting state the nerve of the squid *Loligo* has a value of 0·04 for the ratio permeability to sodium/permeability to potassium. However when the nerve is stimulated the value changes temporarily to 20·0. This change in the permeability of the cell membrane to sodium when stimulated causes the potential across the cell membrane to change in the direction of the sodium equilibrium potential. This sudden change in potential is propagated along the nerve. The permeability to sodium soon decreases and the potential returns to its resting value. The passage of a single propagated action potential results in only a small change in the concentration of sodium and potassium inside the nerve but in the absence of the active extrusion of sodium a series of stimuli would eventually result in an appreciable change in the internal composition of the cell and the nerve would become inexcitable.

The blood of an animal is usually more alkaline than its cells. Thus the pH of the blood of mammals is in the range of 7·35–7·45, whilst that of resting voluntary muscle is about 6·98. A change in the pH of the blood results in changes in the ionic composition of cells. An increase in blood acidity interferes with the operation of the sodium extrusion mechanism and hence there is an increase in the cell sodium level at the expense of potassium. Potassium may also be lost from cells in exchange for hydrogen ions. Variation in intracellular pH, also affects the ionisation of proteins and hence the amount of ions bound.

The rate at which sodium enters a nerve or muscle will vary according to the activity of the excitable tissue. Consequently, the rate of extrusion must also be capable of modification even in animals whose blood concentration does not vary. The capacity to vary the rate at which sodium is extruded from cells is even more important in animals whose blood concentration is liable to fluctuate since the higher the concentration of the blood the greater is the rate of entry of sodium into cells. Penetration of this ion would therefore be expected to be more rapid in marine invertebrates than in terrestrial or fresh water forms. In confirmation of this it is found that when radioactive sodium is placed in the medium bathing isolated muscles from the marine *Mytilus* and the fresh water *Anodonta* it exchanges more rapidly with the tissues of the former than the latter. *Anodonta* is confined to fresh water but *Mytilus* and other brackish water animals may experience considerable fluctuations in the concentration of the blood, and hence in the rate at which sodium enters the cells. Failure to match the rate at which sodium entered would be followed by an outward leakage of potassium, a fall in membrane potential, and entry of chloride. This in turn would lead to failure of cell function.

In fact, it is found that animals living in environments which fluctuate in salinity have the capacity to adjust the regulation of their cells. The crab *Carcinus* experiences considerable changes in its blood concentration and yet its muscle cells maintain an almost constant membrane potential. It is only when the blood potassium is reduced to a low level that the sodium extrusion from the muscles eventually fails to match the sodium inflow. Chloride then penetrates and raises the normally constant Cl cell/Cl blood ratio. Other euryhaline forms such as the amphipod *Gammarus duebeni* and the cave isopod, *Caecosphaeroma burgundum*, also maintain rather constant Cl_c/Cl_b ratios over a wide range of blood concentrations which suggests that their cellular regulation is similarly maintained. On the other hand the cellular ion regulation of both fresh water and marine stenohaline forms breaks down when the blood concentration is varied. Thus, when *Gammarus pulex* from fresh water is placed in saline media, the cell chloride level rises more rapidly than the blood chloride and the animal dies when the salinity of the medium is about the same as the original blood concentration.

The extrusion of sodium is important not only in the maintenance of intra-cellular ion concentration but also in the regulation of cell volume. Recent evidence suggests that most cells are isosmotic with the blood but that they are isotonic only so long as sodium is being extruded as fast as it enters. Poisoned cells not only take up sodium and chloride, they also swell. We might expect that this would be the case. The organic molecules of the cells cannot readily escape and hence they exert a colloid osmotic pressure across the cell membrane. Consequently, even when ionic equilibrium is reached water will continue to move into a cell until prevented by hydrostatic pressure, or until the cell ruptures. It is now assumed that the factor preventing a net movement of water into normally metabolising cells is the extrusion of sodium as it is argued that since sodium is continuously extruded the cell membrane is, in effect, rendered impermeable to this ion. Consequently, the sodium in the blood offsets the colloid osmotic pressure of the organic molecules in the cell. Failure of the sodium pump results in sodium and chloride entry and increase in water content of the cell. If the sodium extrusion mechanism is reactivated after a period of failure the salt can be removed and the volume returns to the normal level. Such an effect can be shown on red blood cells. When these cells are stored at a low temperature sodium extrusion is decreased and the sodium and water content rises. If the cells are then re-warmed to 37° C. in the presence of a suitable respiratory substrate the sodium is largely extruded, potassium regained and the volume returns to normal.

REGULATION OF OSMOTIC CONCENTRATION

Certain organisms have cells which are bathed by media which is hypotonic to the cell fluid, and some alternative method then has to be adopted to remove excess water. The most obvious examples of this situation are seen in the fresh water animals which lack blood systems and have cells directly bathed by the medium. Protozoa, Sponges and Coelenterata come into this category.

Contractile vacuoles

Most of the marine amoeboid protozoa (Rhizopoda) lack any overt means of regulating their water content. It is probable, however, that all are isosmotic with the medium and that regulation is effected in a manner similar to that described above for the cells of higher animals. A number of marine flagellates and most marine ciliates have the capacity to extrude fluid from the cell via a contractile vacuole. Contractile vacuoles are found in all fresh water protozoa and in the cells of fresh water sponges.

It has been claimed that the vacuole serves as a means of removing excretory waste products but its absence in some marine forms and presence in all fresh water species suggests that its main role is concerned with the regulation of cellular volume. This is supported by the observations that (a) many marine protozoa which normally lack vacuoles develop them when placed in dilute media, and (b) when the vacuoles are poisoned with cyanide the protozoa swell. It is probable that those marine protozoa that have vacuoles utilise them for the removal of water taken in during feeding but it is possible that they are also used in the regulation of the ion ratios in the cell.

A measure of the internal osmotic pressure of protozoa has been obtained by determination of the concentration of a non-penetrating substance that must be present in the medium in order to stop cells swelling when poisoned with cyanide. Measured in this way the Suctorian *Podophrya* is found to have an internal concentration of about 40–50 m. osmoles and probably most fresh water protozoa have similar concentrations. The cell membrane of protozoa is less permeable to water than that of most cells of higher organisms but as they are so hypertonic to fresh water a considerable volume of fluid enters by osmosis and has to be expelled. The amount of fluid entering varies with the concentration gradient across the surface so that when *Podophrya* is transferred from fresh water to 0·1 Methylene glycol the body shrinks and the contractile vacuole stops working. After about ½ to 1½ hours the body reswells as glycol penetrates into the cell and vacuolar output is resumed. If the animal is then returned to fresh water its internal concentration is well above normal because of the penetrated glycol. Vacuolar output increases greatly and compensates so successfully for the increased rate of water entry that the cell does not obviously swell above its normal size. As the glycol is eliminated the vacuolar output declines once more. There is usually a slight time lag in the response of the contractile vacuole after the medium has been changed which suggests that it is a change in the volume of the body which is important in initiating the appropriate change in vacuolar response.

There are three possible ways in which water might be caused to enter a vacuole: by pressure filtration through the vacuolar membrane, by osmosis if the vacuole contains a higher solute content than the cytoplasm, and by secretion.

If filtration were important it would be expected that the rate at which vacuoles fill would be associated with the hydrostatic pressure inside the cell. Two factors suggest that this is not the case, (a) certain protozoa have two vacuoles and these alternate in operation, (b) some protozoa form and discharge vacuoles even when the cell is slightly shrunken, though the rate of output is sub-normal. Filtration is also inherently improbable as the formative mechanism, since the hydrostatic pressure in the vacuole is likely to be at the same level as that in the cytoplasm.

Osmosis seems to be eliminated (unless solutes are continuously secreted into the vacuole) because the rate of filling of vacuoles is more or less constant. Despite the increasing surface area of the vacuole as it fills, this could not be expected to occur if the amount of solute inside the vacuole remained constant throughout. It is probable therefore that some form of secretion is involved in filling. What form this secretion may take is as yet unknown as there have been no determinations of the concentration of the fluid in the vacuole. If it is pure water that is expelled from the cell, then it must be water that is secreted into the vacuole. Alternatively, salts and other solutes might be continuously secreted into the vacuole and water enter by osmosis. If the latter is the case, then the vacuolar fluid should be isotonic or hypertonic to the cytoplasm and there will be a loss of salts from the animal each time the vacuole discharges. Such a salt loss would have to be made good by uptake from the medium across the cell wall. Certainly there is a continuous exchange of sodium and potassium between *Spirostomum* and its medium but how the loss of these ions from the body takes place is not yet known.

Cell regulation in coelenterates

Fresh water coelenterates, like the sponges and protozoa, are hypertonic to their medium yet they have no contractile vacuoles. The means by which the cellular water content is regulated is not known.

The brackish-water hydroid, *Cordylophora*, varies in structure according to the concentration of the medium in which it is living. Colonies in fresh water have short stout hydranths born singly from the stolon. Colonies in 15 per cent sea water have longer narrower hydranths carried on branching hydrocauli. As a result of the difference in shape, the surface to volume ratio of the hydranth of the fresh water form is less than that of the brackish-water form. The difference is also reflected at the cellular level. Colonies in fresh water have cells of columnar shape, those in 15 per cent sea water have cuboidal cells (Fig. 17). One effect of this is that a smaller proportion of the surface area of the cells of the animals in fresh water is exposed to the medium than is the case with the others. This may well be important in the water and salt regulation of the colonies in fresh water.

Cell regulation in higher organisms

We have noted that the cells of higher multicellular organisms and perhaps also the marine protozoa and coelenterates regulate their water content by continuously extruding sodium. This will maintain their volume as long as they are isosmotic with the bathing medium. Change in the concentration of the extracellular fluid results however in a net movement of water to or from the cells. If the blood is diluted the cells swell and if it is concentrated they shrink. Changes in the blood concentration of marine, fresh water and most terrestrial animals are usually small and consequently these forms do not require special additional mechanisms to limit the small changes in volume to which their cells may be subjected as a result of variations in the

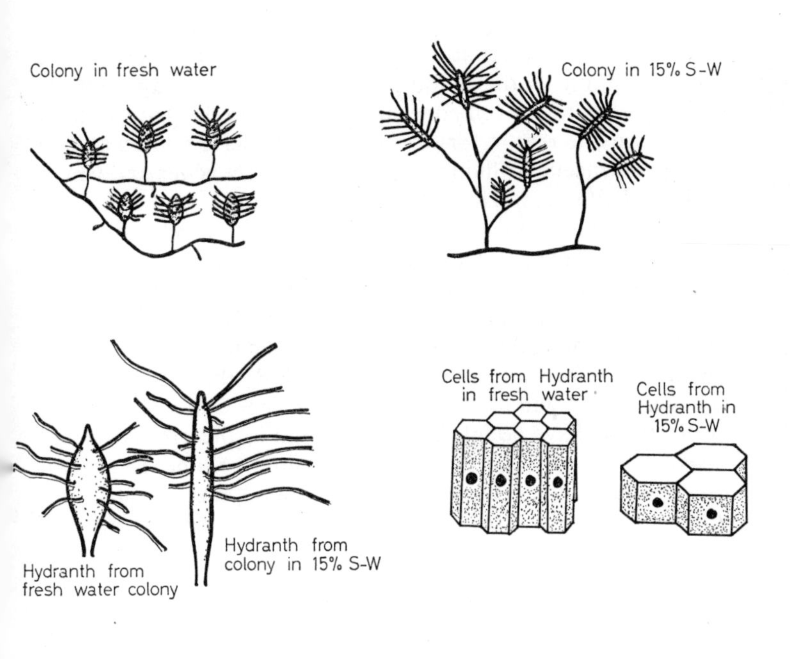

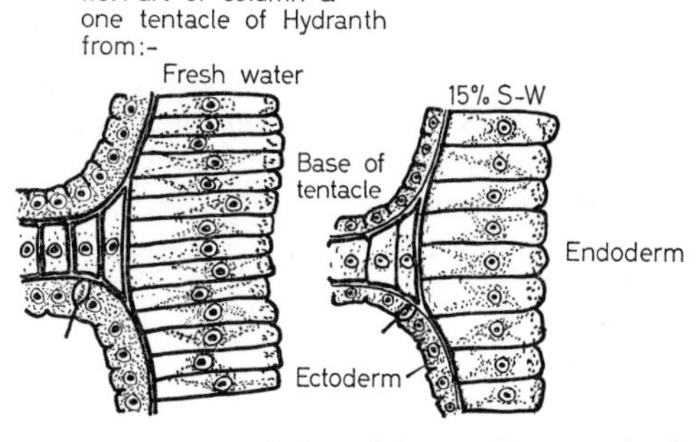

FIG. 17. The effect of the salinity of the medium on the form of the cells and hydranths of *Cordylophora*. (After Kinne)

concentration of the extra-cellular fluid. Brackish water species and those terrestrial forms subject to wide changes in the concentration of the blood must however make appropriate adjustments to the cell concentration. We do not have to look far to see one reason for this. Let us consider a marine animal whose extracellular fluid is isosmotic with sea water and whose cells fill half the total fluid volume of the animal. We will assume that there is no regulation of cell volume, that the cells behave as perfect osmometers, and that the total volume of the animal remains constant. If this hypothetical animal was placed in a dilute medium its blood would be gradually diluted by salt loss and its cells would thus become hypertonic to the blood. They would therefore take up water and swell. By the time the blood concentration had fallen to a value of 50 per cent sea water the cells would have doubled in volume and in the process would have expanded to fill the entire body fluid volume. In general the blood volume would be expected to tend to zero when:

$$\frac{Cb}{Cb_0} = \frac{Vao - Vbo}{Vao}$$

Where C_b is the final concentration of the blood, Cb_0 is the initial concentration of the blood Va_0 is the total fluid volume of the animal and Vb_0 is the initial extracellular volume.

It is obvious that were such changes to take place in any real animal the restriction of the extracellular volume causing circulatory failure or cell rupture would result in death long before the cells expanded to fill the entire animal. Now many invertebrates species have an extracellular volume accounting for 30 per cent or less of the animal's fluid volume and yet as we have seen they tolerate wide changes in blood concentration. In these animals the cell volume is regulated by varying the solute content of the cells as the blood concentration changes. Candidature for the role of regulator substance is rather limited. It might seem possible that the inorganic ion concentration could be varied but in fact as we have seen in the muscles of a marine form such as *Myxine* only about 40 per cent of the osmotic activity of cells is accounted for by inorganic ions. Indeed the ratio of the concentration of ions between blood and medium is rather well maintained as the blood is diluted so changes in their concentration contribute only partly to the change in cell osmotic pressure. Proteins, are also ruled out since, though accounting for much of the total solid in cells, they are responsible for only a minute proportion of the total osmotic pressure. There remain the smaller organic molecules such as amino-acids and trimethyamine oxide. The muscles of marine crustacea, molluscs, hagfishes and presumably also other animals contain a high concentration of these substances. In contrast the amount present in fresh water forms is much smaller. Study of the amino-acid content of the muscles and nerves of *Carcinus* and *Eriocheir* indicate that their concentrations can be varied according to the concentration of the blood. All the amino-acids present seem to be involved in the change but variations in glycine and proline are conspicuously large. The changes may be quite extensive. *Carcinus* loses amino-acids equivalent in concentration to 173 mM/kg from its muscle fibres when the animal is transferred from 100–40 per cent sea water. This decrease in cell solute has the effect of limiting the increase in cell water content to about 5 per cent of its initial value. The isopod, *Caecosphaeroma* is even more successful in regulating the fluid volume of its cells as its muscle water content changes by only 3 per cent when the animal is transferred from fresh water to sea water.

In contrast to *Caecosphaeroma*, most fresh water animals die when the concentration of their medium is raised markedly and this may be attributed in part to their failure to make the necessary adjustments at the cellular level in response to an increase in blood concentration. Fresh water animals are not however totally lacking in the means to vary their amino-acid content. Both the crayfish, *Astacus*, and the crab, *Potamon*, show some increase in amino-acid levels when the concentration of the medium is raised; but they are unable to make sufficiently extensive adjustments to enable them to live in sea water.

The stimulus which brings about a change in the amino-acid level of the cells appears to be associated with a change in the intracellular sodium concentration. When nerves are isolated from the crab *Eriocheir* and are placed in sea water there is an increase in the amino-acid content but when placed in 50 per cent sea water there is no such change. Sugar added to 50 per cent sea water to raise the total concentration to a level equivalent to sea water does not cause an increase in amino-acid, so it is presumed that it is the sodium in the sea water which is responsible for stimulating amino-acid increase.

SUMMARY

Quantitatively the most important inorganic ions of cells are potassium, phosphates and magnesium. Sodium, chloride and calcium are present in lower levels than in the blood. In fresh water and terrestrial organisms inorganic ions may account for over 80 per cent of the osmotic pressure of the cell contents, in marine forms only about 30 per cent to 40 per cent of the osmotic pressure is due to inorganic ions.

The low sodium concentration of cells is maintained by the continuous extrusion of sodium. Cells are isosmotic with the extracellular fluid bathing them but are only isotonic as long as sodium is removed. Failure to extrude sodium results in potassium loss and the entry of sodium, chloride and water.

The rate of sodium extrusion is variable. In brackish water animals this is important as the gradient between cell sodium and blood sodium varies with the concentration of the latter.

Protozoa, sponge cells and coelentrate cells are hypertonic to the medium bathing them. Protozoa and sponge cells extrude water taken up by osmosis through contractile vacuoles. It is not yet known how fresh water coelentrates osmo-regulate. The shape of the cells is different when *Cordylophora* is grown in various salinities.

Brackish water forms compensate for changes in the concentration of the blood by varying the amino-acid level of their cells. When the blood concentration is high so is the cell amino-acid concentration and vice versa.

6

The Active Uptake of Ions

The maintenance of the blood concentration hyperosmotic to the medium results in a loss of ions in the urine and across the body surface. Such losses must be balanced either by the absorption of ions from the food or by direct uptake of ions from the medium. Moreover, as the rates of loss of different types of ion vary, the rates of uptake must be regulated in order to maintain the correct ion ratios in the blood. Such regulation necessitates that the different ion species should be taken up independently of one another.

A few animals, such as the fresh water crustaceans *Daphnia* and *Chirocephalus*, are dependent on a continuous supply of food for the maintenance of their blood concentration. If they are starved the blood concentration falls. It should not be concluded, however, that they rely on food as a source of ions as it may be that in the absence of food the fall in the metabolic rate is such as to decrease the energy supply available for ion uptake at the body surface. Certainly *Chirocephalus* is not incapable of taking up ions directly from the medium as the rate of fall of the body fluid concentration is much more rapid when the animals are in distilled water than when starved in their normal fresh water medium. Most other fresh and brackish water animals can take up ions from the water as fast as they are lost from the body even when they are starved.

Different ion species can be taken up independently of one another. Salt depleted wool-handed crabs (*Eriocheir*) take up chloride from calcium chloride, ammonium chloride, potassium chloride, and sodium chloride, and sodium from sodium chloride and sodium bicarbonate. Except in the case of sodium chloride the other ion of the salt is not taken up. Now, as in the case of the transport of ions by a single cell, uptake of a charged particle by the whole body can only occur if an ion of opposite charge is taken up or, alternatively, if an ion of the same charge is lost. Thus the uptake of sodium or chloride without an accompanying counter ion necessitates simultaneous loss of charge from the body. It appears that sodium uptake is at the expense of ammonium ions or possibly hydrogen ions and that bicarbonate is liberated when chloride is the ion being actively taken in. Chloride and sodium are not invariably taken up by forced exchange for bicarbonate and ammonium. When sodium and chloride are being taken up simultaneously by the gills of *Eriocheir* the conductivity of the medium falls, indicating that the ions taken in are not substituted by ammonium bicarbonate.

The factors which affect the rate at which ions are taken up from the medium have been investigated on the crayfish, *Astacus*.* The two factors of major importance in regulating sodium uptake rate are (a) the blood concentration and (b) the medium concentration. (a) When crayfish are placed in distilled water the sodium ions lost in the urine and by diffusion naturally cannot be replaced. The blood concentration therefore falls. This fall in blood concentration activates the mechanisms responsible for ion transport so that when the crayfish is returned to fresh water the rate of uptake of sodium is above the normal rate and exceeds that required to balance the rate of loss of this ion from the body. The blood concentration therefore rises once more. A lowering of the blood concentration by only 1 to 2 per cent produces some increase in the rate of sodium uptake but the maximum rate of uptake is not reached until the blood concentration is some 6 to 8 per cent below normal. The rate of uptake is therefore linked to the concentration of the blood, and the blood concentration comes to a steady state when loss and uptake of ions are equal. (b) If the concentration of the medium is lowered there is no change in the rate of sodium uptake until a concentration of about 1 mM/l NaCl is reached. Below this concentration the rate of uptake is related to the concentration

* The generic name of the species used is now, *Austropotamobius*.

of the medium, declining more rapidly the greater the dilution. The shape of the curve relating the concentration of the medium and the rate of sodium uptake bears a marked similarity to the Michaelis-Menton curve which defines the reaction between enzymes and their substrates (Fig. 18). One possible explanation

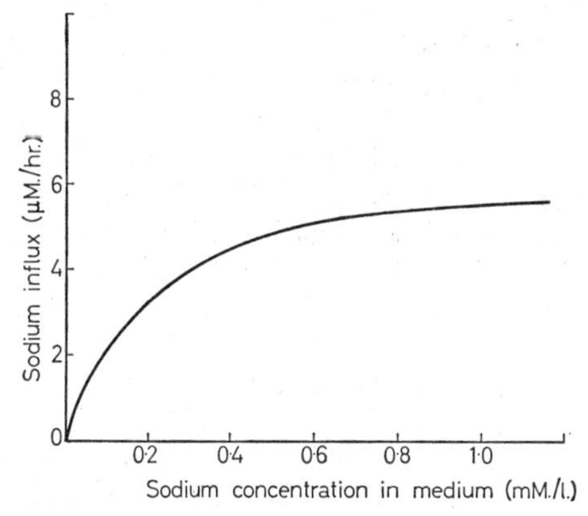

FIG. 18. The relationship between the concentration of the medium and the rate at which sodium is taken up by salt-depleted crayfish. (After Shaw)

for the curve having this shape would be that the active uptake of sodium involves a stage in which sodium is linked to some carrier molecule. At concentrations of the medium below that required to saturate this carrier the rate of uptake would be below normal.

As a consequence of the decrease in the rate of uptake in media less concentrated than 1 mM/l the diffusion and urine loss of sodium is no longer offset and the blood concentration falls below its normal value. This does not mean, however, that the crayfish is unable to live in water less concentrated than 1 mM/l for, as noted above, a fall in blood concentration increases the rate of uptake. Hence, if the rate of uptake were to be initially halved by a change in the concentration of the medium, then the blood would come to a new steady state at the concentration at which the rate of uptake would have been doubled had the medium remained more concentrated than 1 mM/l. At this point loss and uptake are again the same.

The interaction of rate of loss, blood concentration and medium concentration is shown for a hypothetical animal in Fig. 19. Line X represents the rate of uptake at different blood concentrations when the transporting mechanism is fully saturated. Line Y is the rate of uptake when the animal is in a medium which only half saturates the transporting system. Line R is the normal rate of loss at different blood concentrations. Q is the loss if the surface permeability is twice normal. (For the purpose of drawing lines Q and R it is assumed that the urine concentration does not vary when the blood concentration is changed and that the permeability remains constant.) Line S is the loss curve when the concentration of the medium is half that of the animal's blood. Line X cuts line R at a and the 'Normal' blood is at A. If the medium concentration is lowered until the transporting system is only half saturated the blood concentration is not also halved but falls only to B since lines Y and R cut at d. The same blood level is present if the permeability is doubled in a more concentrated medium, curves X and Q cutting at b. The effect of changes in permeability or saturation of the transporting system on the blood concentration will thus depend on the steepness of the slope of the curves X and Y at the normal blood level.

Raising the concentration of the medium also has rather little effect on the blood concentration until levels equivalent to the normal blood concentration are reached. In the case illustrated in Fig. 19 the blood concentration rises to C, only some 7 per cent above normal, when the medium concentration is raised from normal (fresh water) to a level equivalent to half the blood concentration.

The concentration of the medium required to saturate the active transport mechanisms must obviously affect the environments which can be colonised by an animal. We should expect that species whose transport mechanisms are only fully saturated when the concentration of the medium is high would

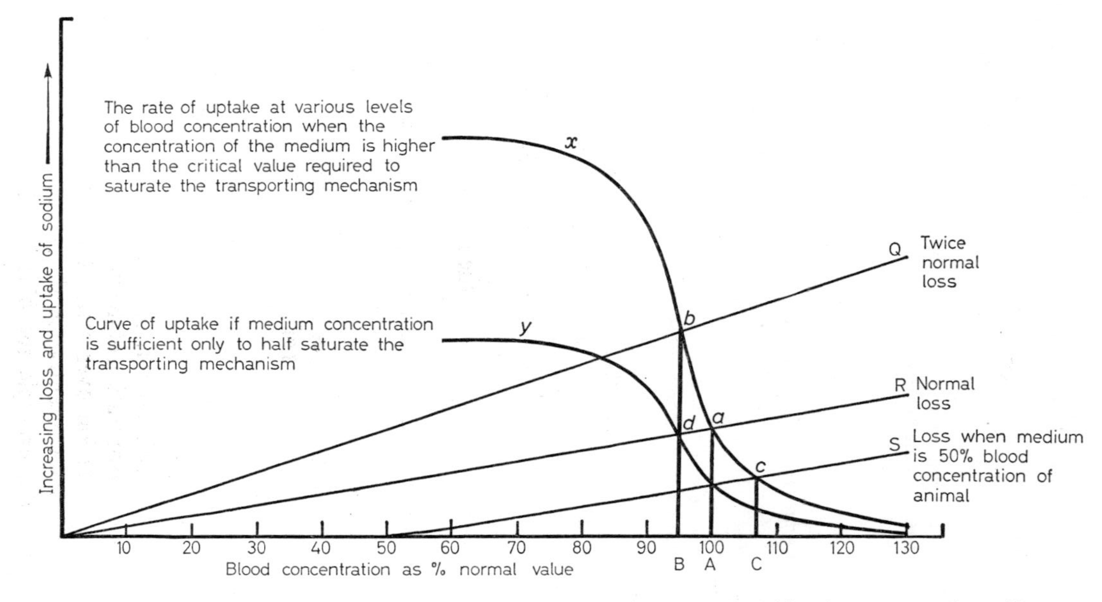

FIG. 19. The interaction of various factors involved in the control of blood concentration. (See text for explanation)

be unable to colonise dilute media, and hence that forms which do live in dilute media would have evolved lower levels for saturation than marine and brackish water forms. It has been shown that this is the case for Crustacea (Table 17). The

TABLE 17. The concentration of sodium necessary in the medium in order to half saturate the sodium transport mechanisms of animals from habitats of various salinities. (Data from Shaw)

Species	*Normal medium*	*Conc. for half saturation*
Carcinus maenas	Sea water to brackish water	20 mE/1.
Gammarus duebeni	Brackish to fresh water	1·5 mE/l.
Astacus pallipes	Fresh water	0·3 mE/l.
Gammarus pulex	Fresh water	0·15 mE/l.

level at which the transporting mechanisms are saturated is not necessarily the same for all ions. Thus, for the crayfish, the medium concentration required to saturate the mechanism responsible for chloride transport is only 0·2–0·6 mM/l, as opposed to 0·5–1·0 for sodium. Similarly the saturation levels for sodium and potassium differ in the East African fresh water crab *Potamon niloticus*.

An increase in the concentration of the medium above that necessary to saturate the transporting mechanisms does not affect the rate of active uptake. Consequently, a rise in the concentration of the medium does not affect the concentration of the blood until the rate of loss of ions from the body decreases as the concentration gradient falls. The blood concentration then rises. Brackish-water animals can compensate at the cellular level for such changes in blood concentration (p. 99) but most fresh water animals die when the concentration of the medium approximates to their normal blood concentration. Nevertheless, because of the operation of the system shown in Fig. 19, the range of concentration over which fresh water animals may maintain the blood concentration almost constant may be very considerable. For example the blood of the isopod *Asellus aquaticus* varies in concentration by only 10 to 15 mM/l when the animals are in media ranging in concentration from 0·1 to 60 mM/l, a 600 fold concentration difference.

Temperature is another factor influencing the rate of active uptake by poikilothermic animals. When the temperature is decreased the rate of active uptake falls. In the case of *Asellus* the rate of ion loss from the body is not affected by temperature so any change results in an imbalance between uptake and loss of ions. However, the fall in blood concentration following a drop in temperature activates the uptake mechanism as before so that a new steady state is established at a somewhat lower blood concentration. Temperature has a similar effect on the active uptake by the lamprey, *Petromyzon* but it is possible that in some other animals it will be found that temperature change affects the rate of loss more than the rate of uptake.

The site of active uptake from the medium

Ions are taken up by the whole body surface of frogs and this is probably also the case in many other soft-bodied forms such as annelids and molluscs. Frogs show some localisation of uptake however in so far as the skin of the belly transports sodium more rapidly than does that of the back. Fish and hard-bodied animals such as the arthropods restrict the permeability of the general body surface and ion transport is confined to the one area with a high permeability, the gills.

Conclusive evidence has been provided of the part played by the gills of crustacea in the uptake of ions. Isolated gills of *Eriocheir* suspended in well oxygenated 8 mM/l NaCl take up sodium from the medium even though the blood in the gills is some thirty times more concentrated than the medium. When the gills are poisoned with carbon dioxide or anti-cholinesterases uptake ceases and there is a net loss of sodium from the gills. Uptake begins again when the poisoning agent is removed. Lack of oxygen also results in a net loss of sodium. These features indicate that the uptake of sodium is an active process and that the rate of uptake can more than compensate for the normal loss of sodium that occurs continuously across the surface of the gill at the same time as uptake is proceeding. These gills also take up potassium and the fact that there is no change in the rate of uptake of sodium while potassium is being taken in indicates that the mechanisms responsible for the transport of the two ions are independent.

The cuttlefish buoyancy mechanism

So far we have considered active transport only as a means of adjusting the concentration of blood or cells. The cuttlefish *Sepia*, however, utilises the active regulation of the concentration of the fluid inside its cuttlebone in order to adjust its buoyancy. The cuttlebone is composed of a large number of chambers separated by lamellae of calcified chitin. The lamellae are separated from each other by pillar-like struts and the whole cuttlebone has considerable mechanical strength. In life the chambers are partly filled with gas and partly with liquid, the latter occupying the lower part of the chamber which is in contact with a living epithelium. The presence of the gas in the cuttlebone offsets the tendency of the animal to sink and gives it practically neutral buoyancy. Hydrostatic pressure varies with depth. Hence, it would be expected that as the animal swims downwards in the water the gas would be compressed and the proportion of fluid in the chamber increased. If the gas were to be compressed in this manner the density of the animal would increase the deeper it went and hence the more effort it would require to swim. *Sepia* overcomes this difficulty by offsetting osmotic and hydrostatic pressures in rather a novel manner. When the animal is exposed to pressures of several atmospheres, the concentration of the fluid in the cuttlebone is actively decreased. The tendency for fluid to be forced through the bounding membrane into the cuttlebone is thus offset by the equal tendency of the greater osmotic pressure of the body fluids to withdraw water from the chamber. The gas volume thus remains unchanged and so does the buoyancy.

The liquid in the chamber resembles the blood in that a high proportion of its osmotic activity is provided by sodium and chloride, and changes in the concentration of these ions are primarily responsible for the variations in osmotic pressure. It is not known precisely how the change in pressure is effected but it seems likely that sodium and chloride are actively transported from chamber to blood by the bounding epithelial cells. The extent to which the chamber fluid can be diluted is uncertain but the theoretical limit for the mechanism would obviously be reached when it was reduced to pure water. As the blood of *Sepia* is isosmotic with sea water it has an osmotic pressure approximately equivalent to a one molar solution of non-electrolyte. The maximum theoretical gradient that could be sustained across the bounding epithelium would thus be 1 mol or 22·4 atmospheres. At a depth of 750 feet the hydrostatic pressure would be approximately equivalent to 22·4 atmospheres and this is the maximum depth at which the gas volume could be kept constant. *Sepia* is not known to penetrate to so great a depth, though it is thought to go as deep as 600 ft.

Deep water cephalopods cannot use this mechanism. Instead some of them concentrate ammonium chloride in the coelomic spaces. Presumably the concentration of this substance also involves the transport of sodium out of the coelomic fluid. Ammonium chloride is less dense than sodium chloride and therefore serves as a float. However, so large a volume is required to be effective that these forms are usually very watery.

Possible means of active transport

Before considering the possible means by which cells transport ions we must first define what is meant by the term active transport. Ions can be moved passively by electrical gradients in a direction opposed to that of the concentration gradient. Hence an ion can only be definitely said to be actively transported when it is moved against the direction in which the combined electrical and concentration gradients would normally take it. Now it is obvious that if there is a potential across a membrane it will tend to move positively charged ions one way and negatively charged ions the other. Any potential difference therefore increases the gradient through which ions of one charge must be moved but facilitates the movement of oppositely charged particles. This effect has been shown on the frog skin. The frog skin has a potential across it such that the inside is positive to the outside. This potential would be expected to aid the entry of chloride and increase the minimum work necessary to transport sodium. It has been shown that when the potential difference across the skin is artificially balanced chloride uptake ceases but that of sodium continues. It is concluded that the skin actively transports sodium but that the chloride normally moves passively down the electro-chemical gradient. The electrical potential necessary to balance a chemical gradient of an ion is given by the formula

$$E = \frac{RT}{zF} \ln \frac{C_i}{C_o}$$

where E is the potential, R is the gas constant, T is the absolute temperature, and C_i and C_o are the concentrations on either side. z is the valency of the ion and F the Faraday.

The extrusion of sodium by nerves, muscles and other cells is dependent on the presence of an adequate concentration of potassium in the medium bathing the cell. If the potassium is at too low a concentration sodium transport is decreased. The frog skin is similarly sensitive to low potassium levels in the medium bathing the inner surface of the skin and it has been suggested that the active transport mechanism taking up sodium from the medium may be a modification of the mechanism used in the sodium regulation of the general body cells. Transport of ions from the medium to the extracellular fluids involves the movement of ions across at least the two cell membranes of an epithelial cell. Obviously both these membranes cannot be transporting sodium outwards from the cell, one at least must be modified from the condition found in nerves. It is possible that the inward facing membrane behaves like a resting nerve cell membrane, almost impermeable to the passive movement of sodium but actively transporting sodium from the cell to the blood in exchange for potassium. The outer cell membrane may be more like a nerve membrane during the action potential, very much more permeable to sodium than to potassium. The series of events involved in the transport of sodium and chloride ions from medium to blood is assumed to occur as follows. The sodium in the cell is initially in electro-chemical equilibrium with that in the medium. When sodium is transported from the cell into the blood the electrochemical potential of sodium inside the cell declines and sodium diffuses down the gradient from the medium through the sodium permeable external cell wall. Potassium taken into the cell from the blood in exchange for the extruded sodium can leak back into the blood down its chemical gradient provided that it is accompanied by an ion of opposite charge. Chloride moving down the electrochemical gradient provides the negative charge. Diffusion of potassium and chloride into the blood completes the cycle. The initial state is now regained except that sodium and chloride have been transported from medium to blood. This ingenious hypothesis

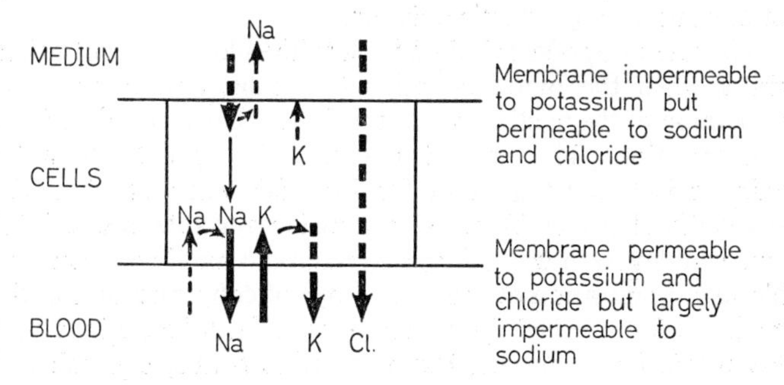

FIG. 20. Theoretical model of the active transport of sodium chloride across an epithelial cell. (See text for explanation)

accounts for all the experimental results obtained on the frog skin but it may not be universally applicable to sodium and chloride transport. For instance, it cannot account for the uptake of sodium chloride in the crustacea. In these animals the polarity of the gill is the reverse of that of the frog skin, the inside of the animal now being negative rather than positive to the outside. Chloride will not diffuse passively into the animal in such circumstances; it must be actively transported. When a frog has a low blood chloride it too can take up chloride actively. Possibly the difference between the transport in crustacea and the frog lies in differences in the overall permeability of the transporting epithelium to cations and anions.

In the air breathing vertebrates uptake of ions through the skin is no longer possible and these animals and other terrestrial forms take in all their ions through the gut. Such intake may be at infrequent intervals and hence regulation of the body fluid composition devolves largely on the control by the kidney of the relative rates of loss of ions and water.

SUMMARY

Various ions such as sodium, potassium, chloride etc. can be taken up against the electrochemical gradient independently of one another by aquatic animals.

The rate of uptake is linked both to the blood concentration, rising when the blood concentration falls, and to the concentration of the medium when this is less than a certain critical concentration. The critical concentration for fresh water animals is lower than that for brackish water species. The association between blood concentration and rate of uptake of ions enables fresh water forms to maintain an almost constant blood concentration over a wide range of concentrations of the medium. Soft bodied forms, e.g. amphibia, take up ions over the whole body surface, less permeable forms, e.g. fish and crustacea take up ions at the gills.

The mechanism responsible for the uptake of sodium appears to be based on the mechanism utilised to extrude sodium from the general cells of the body.

7

The Role of the Kidney

VERTEBRATES

The kidneys of vertebrates are responsible in whole or in part for five aspects of body fluid regulation. These include:

(1) Removal of metabolic waste products and foreign molecules.
(2) Regulation of the osmotic concentration of the body fluids.
(3) Regulation of the concentrations of different ions.
(4) Regulation of the reaction (pH) of the blood.
(5) Regulation of the volume of the body fluids.

The capacity of the kidney to perform these functions is intimately related to its anatomical structure.

Kidney structure

The functional unit of the kidney is the *nephron* and its associated blood supply.

A generalised vertebrate nephron is composed of a cup-shaped *Bowman's capsule* the hollow lumen of which opens into the *renal tubule.* Various regions are generally recognisable in the tubule. Continuous with the neck of Bowman's capsule is the *proximal convoluted tubule.* This is separated from the *distal convoluted tubule* by an *intermediate section.* The distal convoluted tubule is contiguous with a collecting duct which, joining with collecting ducts from other tubules, eventually opens into the *ureter.*

The cup of each Bowman's capsule encloses a mass of capillaries, the *glomerulus*, which receives blood from the *renal artery* via *afferent arterioles.* The glomerulus plus the Bowman's capsule is termed a *Malpighian corpuscle.* Blood leaving the glomerulus is gathered into an *efferent arteriole* which subsequently gives rise to a capillary network enclosing the renal tubule (Fig. 21). In all vertebrates except the cyclostomes (lampreys and hagfishes), birds and mammals the capillary network round the tubules is also supplied with venous blood derived from the tail and hind limbs. Blood from this latter *renal portal* system never supplies the glomeruli.

Each kidney is composed of a large number of nephrons. Man has about one million nephrons in each kidney with a length of 2–4 cms each so that in the two kidneys there is a total of between 55 and 110 miles of tubule. As the diameter of the tubules is small a considerable surface area is made available for the exchange of valuable ions and metabolites.

Kidney function in relation to structure

The primary urine is formed in most vertebrates by a process of ultra-filtration. The hydrostatic pressure of blood entering the glomerulus is considerably in excess of the combined colloid osmotic pressure of the plasma proteins and hydrostatic pressure of the fluid in Bowman's capsule. The membranes separating the blood from the capsule lumen are very thin and readily permeable to small molecules. An ultrafiltrate of blood is therefore forced into the lumen of the capsule. If the renal artery is constricted so as to cause the hydrostatic pressure in the glomerulus to fall urine formation stops.

The hydrostatic pressure in the lumen of the capsule required to force fluid along the tubule at a given rate will vary with the tubule length if other tubule dimensions are constant. A large surface area of cells must be presented to the fluid in the lumen so there is a limit to the possible diameter of the tubules. As a consequence of these two factors large animals tend to increase their renal capacity mainly by increasing the number of nephrons rather than by a great increase in the length of the individual nephrons. Thus whereas a dog has about 800,000 glomeruli and tubules with an average length of about 3·7 cm, a cow has 8,000,000 glomeruli with tubules whose average length is 5·4 cm.

Analyses of the primary urine in the lumen of the capsule

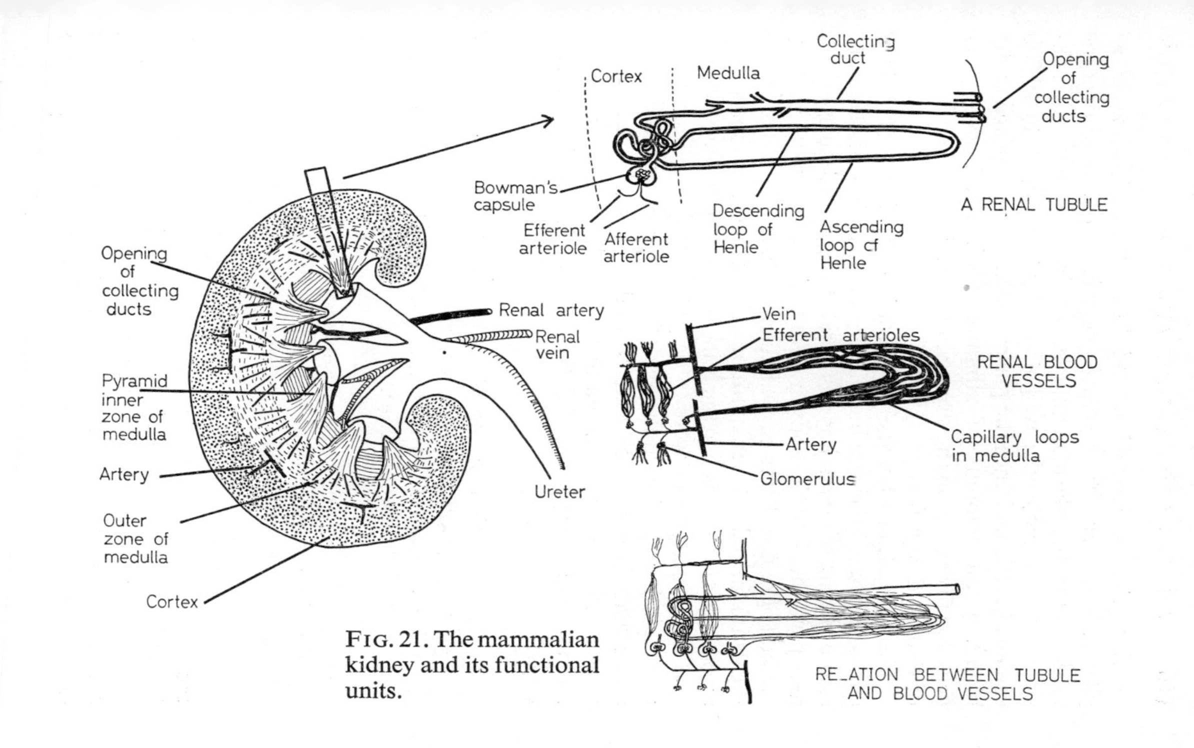

FIG. 21. The mammalian kidney and its functional units.

provide evidence that all ions and the smaller molecules such as glucose and urea are present in effectively the same concentration as in the plasma. Experiments indicate that only substances with a molecular weight smaller than 60,000 to 70,000 can be filtered and hence the plasma proteins are retained in the blood. Small amounts of albumin (Mol. Wt. 70,000) may occasionally escape in the urine but any major loss is indicative of damaged capsules.

The formation of urine by filtration provides not only a convenient way of removing waste products of metabolism but also a means of removing any small foreign molecules which may penetrate into the body and for which no specific secretory method of removal may be available. For example, although the sugars inulin and xylose cannot be metabolised by animals they are filtered and removed in the urine if injected into the blood.

If the primary urine were to be passed unchanged down the renal tubule and released the loss of valuable water, electrolytes and metabolites such as glucose and amino-acids would be large relative to the waste matter removed. Excessive waste of these useful components of the plasma is avoided in two ways:

(1) Concentration of the filtered waste by reabsorption of water, salts and metabolites.
(2) Active secretion of waste products into the urine through the tubular epithelium.

A combination of both these methods is probably always employed but some groups of vertebrates place more emphasis on one than the other. Mammals have a high filtration rate and tend to concentrate their principal nitrogenous waste product urea, by reabsorption of water. Both the volume of filtrate and the proportion of water reabsorbed by the kidneys is much lower in other groups and then secretion is often utilised to concentrate waste products in the urine. Some fish and the frog secrete urea into the tubule whilst birds and terrestrial reptiles secrete uric acid. The capacity of the tubules to secrete is rather variable in mammals. Creatinine is actively secreted by the tubule cells of man as well as by many fish though apparently not by dogs, cats and rabbits. Urea is secreted by some mammals including the desert rat. In addition both dog and human kidneys secrete a variety of other substances including some dyes, hippuric acid, diodrast and, unfortunately, penicillin.

A number of marine teleosts (angler fish, etc.) have nephrons which lack Bowman's capsules and glomeruli. Their tubules are supplied only with blood from the renal portal system, which, being venous, has a low hydrostatic pressure. Urine formation in these forms is believed to depend on a process of secretion since if inulin is injected into the blood it does not appear in the urine. The elimination of filtration by angler fish appears to be associated with restriction of urine volume. Animals living in dry places also tend to reduce the glomerular size. Thus many reptiles have small Bowman's capsules and in some the glomerulus has a non vascular core. The desert frog, *Chiroleptes*, has glomeruli when young but later loses them.

The concentration of the urine

All vertebrates, except the hagfish, can produce urine less concentrated than the blood. The basic processes involved in the production of hyposmotic urine are (1) filtration of fluid isosmotic with the plasma, and (2) reabsorption from the urine as it passes down the renal tubule. Some water usually accompanies the reabsorbed solutes (Table 18) but many fresh water vertebrates can produce very dilute urine. Marine teleosts and terrestrial vertebrates are exposed to the opposite hazard, that of dehydration, and hence need to conserve water at the expense of salts. Apparently, however, the direction of salt transport by the kidney tubules cannot be reversed in order to increase the concentration of salt in the urine above that of the plasma, and marine teleosts, reptiles, amphibia, and elasmobranchs all produce urine which is hyposmotic or isosmotic to the blood. Only birds and mammals can produce urine hyperosmotic to the blood. Even in these latter two groups recent work suggests that the concentration of the urine is achieved by a modification in the structural relations of the renal tubules, collecting ducts and blood vessels rather than by a change in the direction of salt transport.

TABLE 18. The approximate proportion of the volume filtered which is reabsorbed during passage down the renal tubule in various groups of vertebrates.

Animal	*Filtration rate ccs/Kg. body wt./Hr.*	*Urine flow ccs/Kg./Hr.*	*Max. % water reabsorbed*
Teleost			
F–W Catfish	10–18	6–14	48
Amphibian			
Frog	3–40	2–21	68
Reptile			
Lizard	0·3–3·8	0·2–1·2	80
Bird			
Hen	24–92	3–8	94
Mammal			
Dog	92–327	3–19	98

Note (a) the low rate of filtration in the reptile, (b) the high filtration rate in the mammal and (c) the extensive reabsorption of water in the mammals and birds with lesser amounts in the lower groups. (Modified from Marshall)

In the lower vertebrates (fish, amphibia and reptiles) the renal tubules are rather short (except in the elasmobranchs) and are not arranged in any obviously ordered manner within the kidney. In bird kidney there is a greater degree of organisation and two regions can be made out, a small central *medulla* which contains only collecting ducts, blood vessels and some long straight sections of tubule and an outer *cortical* region partially enclosing the medulla, which contains all the glomeruli and the major part of the tubules.

The renal tubules of the outer part of the cortical region are unspecialised and resemble the type found in reptiles. The tubules with glomeruli nearer the junction of the cortex and medulla are more like those of mammals. As in the mammals the proximal and distal regions of the bird tubules lie in the cortical region and undergo tortuous convolutions in the vicinity of their capsules. Separating the two convoluted tubules is a long straight section of tubule which penetrates into the medulla between the collecting ducts and then, looping on itself, runs back into the cortex. In both birds and mammals this section of the tubule is called the *loop of Henle*.

In most bird kidneys tubules with long loops of Henle are in a minority and most of the tubules are of the reptilian type or of an intermediate short-looped type.

The division into cortical and medullary regions is more pronounced in the mammalian kidney. In this group the cortex is usually undivided and forms a cap over a well developed medulla. The latter is subdivided in many species to form a number of *pyramids* at whose tip the main urinary collecting ducts open into the expanded mouth of the ureter (Fig. 21). As in the birds, the glomeruli are confined to the renal cortex and tubules whose capsules lie near the outer part of the cortex tend to have shorter loops between the proximal and distal tubules than those whose capsules are nearer the medulla. The loops of the latter are markedly longer than those in birds and penetrate to near the tip of the pyramids.

The presence of the loop of Henle only in the two groups of vertebrates able to form hyperosmotic urine suggests that this region of the tubule is itself implicated in the process of concentration. This probability is further emphasised by the fact that the proportion of tubules with long ducts is much greater in mammals living in dry habitats than in fresh water forms. The beaver has only short loops of Henle and a maximum urine concentration of about 600 m. osmoles, the rabbit has both short and long tubules and a maximum urine concentration of 1,500 m. osmoles whilst the desert rat, *Psammomys obesus* has only long loops and a maximum concentration of nearly 6,000 m. osmoles. Not only is the proportion of long to short tubules related to the urine concentration but so is the relative thickness of the renal cortex and medulla. In forms with low maximum urine concentration the medulla is very much thinner and smaller than it is in species which can produce very concentrated urine. This can be seen by comparing the size of the medullary pyramid in rodents from aquatic, general terrestrial and desert environments, (Fig. 22).

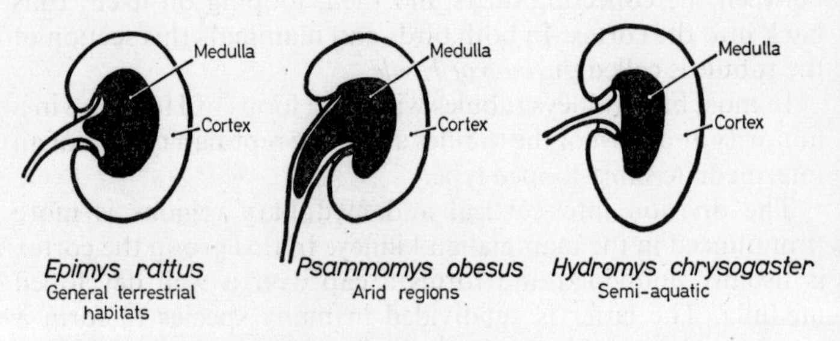

FIG. 22. The kidneys of rodents from various environments to show the relationship between the size of the medulla and the availability of water in the habitat. (Modified after Sperber)

The manner in which the mammalian and bird kidneys produce hyperosmotic urine is believed to depend on a process somewhat analagous to the heat exchangers used by chemical engineers to avoid heat loss from a system. This utilises the counter current principle.

The counter current principle

If two pipes carrying respectively hot fluid and cold fluid are laid in contact with each other heat is transferred and if the direction of flow in the two pipes is opposite then along the area of contact there is a graded decrease in the temperature of the hot pipe and a corresponding increase in the temperature of the other. If the system is a closed one (Fig. 23), and the counter current transfer is effective then there will be only a slow escape of heat. Little energy then need be supplied at the junction of the pipes to maintain a constant temperature. If no counter current exchange is present (Fig. 23) more energy has to be supplied to maintain the temperature.

Consider now the loops of Henle and their associated blood vessels. Their looped shape gives them the structure required

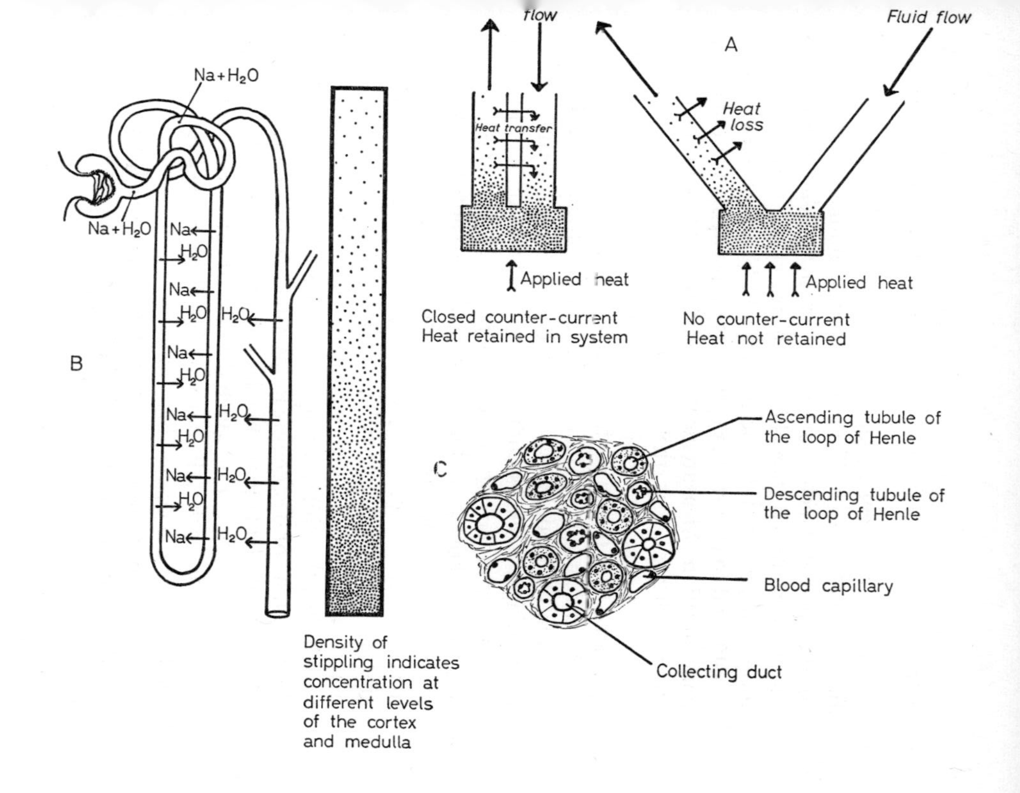

FIG. 23

(A) The counter current principle.

(B) The kidney counter current. To show the movements of sodium and water believed to occur in different regions of the nephron.

(C) Semi-diagrammatic representation of a transverse section through the renal medulla to show the relationship between the tubules of the loop of Henle, the collecting ducts and blood vessels.

in a counter current exchange, not in this case of temperature but of concentration. If a concentration gradient were present in the medulla such that the lower pyramidal region were more concentrated than the region nearer the cortex then the looped arrangement of the tubules and blood vessels (Fig. 23) would tend to ensure that there was no rapid dissipation of the concentration. Measurements of the freezing point of kidney slices indicates that there is such concentration difference between regions. The blood and urine are both more concentrated near the tip of the pyramid than at the junction of the cortex and medulla. This effect could be produced and maintained in the medulla if water were to be transported from the descending to the ascending loop of Henle or if salts were transported in the opposite direction. The latter seems the most likely as the hormone aldosterone which is known to affect salt transport can increase the concentration of the urine. The maximum concentration gradient that can be maintained across any one region of tubule wall is small; but if this small gradient is maintained between the two tubules at all points along their length then the difference in concentration between the ends of the ascending or descending loops will be considerable (Fig. 23). The loop of Henle may thus be regarded as a counter current multiplier, the factor by which the tubular urine is concentrated by the time it reaches the bottom of the loop being a function of the length of the descending tubule.

The concentration depends not only on the length of the tubule but also on the rate of fluid flow. If the urine flow is rapid, then less time is available for the establishment of the gradients between ascending and descending tubules and the maximum concentration at the pyramid tip falls. The ascending and descending loops of Henle are not in direct contact (Fig. 23). Transport of salt or water must therefore be via the interstitial fluid of the medulla and the blood vessels. The rate of blood flow through the medulla will therefore also affect the concentration of the urine. However, the capillary vessels of the medulla, although rather large, carry only about one per cent of the blood that passes through the kidney. Consequently, the rate of flow in these vessels is rather slow. As the blood vessels are also arranged in a loop (Fig. 23) there is little tendency for the blood flow to dissipate the high medullary concentration.

On leaving the ascending tubule of the loop of Henle the urine is passed into the distal convoluted tubule – and thence into the collecting ducts. Now it is found that the concentration of the urine in the distal convoluted tubules is less than that in the blood even when animals are producing definitive urine more concentrated than the blood. It is clear, therefore, that the final concentrating of the urine must take place in the collecting ducts. Control of the final concentration is achieved by a hormone, the antidiuretic hormone (vasopressin, A.D.H.) which is released into the blood from the posterior region of the pituitary gland (see p. 140). When this hormone is present in the blood the cells lining the distal convoluted tubule and collecting ducts become more permeable to water. When little hormone is present the urine flows unchanged in concentration along the collecting ducts to the ureter. When much hormone is present and the tubule cells are permeable, water is withdrawn by osmosis from the distal tubules thus decreasing the urine volume entering the collecting ducts. As the urine then passes down the collecting ducts it is only separated from the high concentration in the medullary fluids by the (now) water permeable duct cells. More water is therefore withdrawn by osmosis and the collecting duct urine assumes a concentration dictated by the maximum concentration in the medullary pyramid, and the rate of flow in the collecting ducts. Thus the function of the high concentration maintained in the medulla is to determine the final concentration of the urine. The effect of A.D.H. is both to concentrate the urine and decrease its volume. During the production of large amounts of very dilute urine the concentration in the medulla may fall (Fig. 24).

In man about 170 litres of fluid (or some 30 times the blood volume) are filtered by the glomeruli per day but only about 1,500 ccs of urine are produced. Thus some 99 per cent of the filtrate is reabsorbed. However, this reabsorption is not all regulated by the antidiuretic hormone. Complete absence of

A.D.H., such as may occur if the pituitary is damaged, only increases the output of urine to about 15–20 litres of urine per day. Facultative control of the urine volume is exercised only over this 10–20 per cent of the initial filtrate. Most of the remaining 80–90 per cent is reabsorbed in the proximal tubule

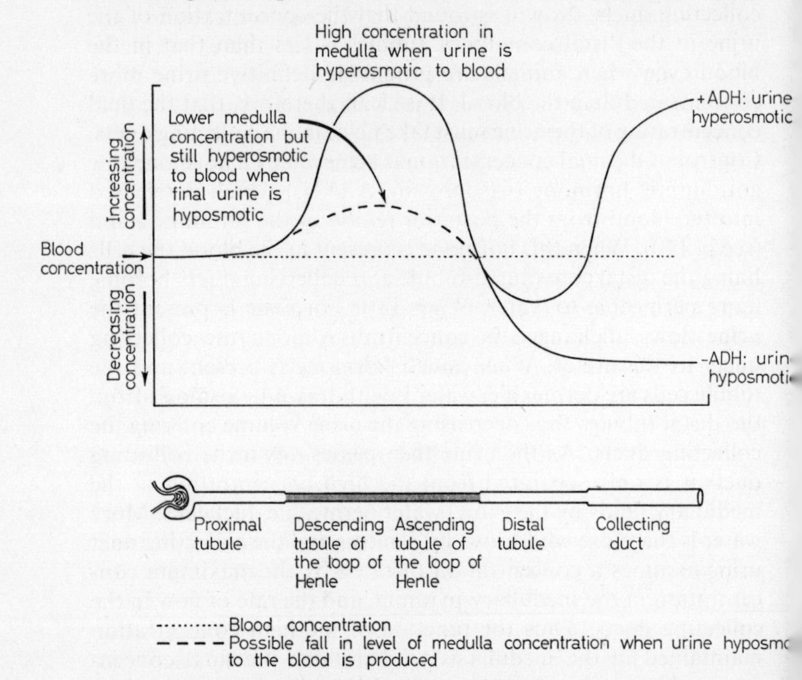

FIG. 24. Diagrammatic representation of the probable concentration of the urine in various parts of the nephron both in the presence and absence of A.D.H.

as a solution isotonic with the blood. Since 20 per cent of the plasma entering the glomeruli is filtered the plasma proteins are correspondingly concentrated in the remaining 80 per cent which passes into the efferent arterioles. These efferent arterioles enclose the convoluted tubules and it has been suggested that the withdrawal of fluid from the proximal tubule is brought about by the high colloid osmotic pressure in the blood they contain. For this to be possible the proximal tubule would have to be very permeable to the major ions in the primary urine. The use of radio-active sodium indicates that the wall of the proximal tubule is indeed rather permeable to sodium, at least much more so than the distal tubule, and it has been calculated that the colloid osmotic pressure is adequate to account for the fluid withdrawal. However, there are three reasons for supposing that an active process rather than colloid osmotic pressure is really involved in the reabsorption of sodium chloride and water in the proximal tubule. These are:

(1) Uptake is much decreased by drugs which are known to poison the active transport of sodium.

(2) Uptake continues, though at a diminished rate, when sufficient albumin is introduced into the lumen of the proximal tubule to offset the colloid osmotic pressure of the blood.

(3) The oxygen requirements of the kidney are related to the passage of fluids.

It is probable that the active transport of sodium provides the drive and that water moves passively.

Useful substances in the initial filtrate such as glucose, phosphates, potassium, vitamin C, and amino acids are also partially or completely reabsorbed in the proximal tubule. Such substances are actively reabsorbed but there is a limit to the rate at which the mechanisms responsible can operate. If the amount of substance to be reabsorbed is present in high concentration in the plasma or the rate of urine flow in the renal tubule is excessively fast reabsorption is incomplete and some will escape into the definitive urine. For example the maximum rate at which glucose is reabsorbed in man is about 350 mgms/min. As the normal filtrate is about 130 ccs/min. and the normal plasma concentration about 100 mgm/100 ccs there is a considerable safety factor present. The maximum rate at which a substance can be reabsorbed is called the tubular maximum or

Tm. In diabetes mellitus the plasma levels of glucose rise and the Tm for glucose may be exceeded. The loss is then proportional to the plasma concentration.

In the case of man the filtration rate is usually rather constant so it is only variations in the plasma concentration that are likely to result in the tubular maximum for any substance being exceeded. Many of the lower vertebrates have more labile filtration rates and rate of flow may then be important.

Inulin is neither metabolised nor reabsorbed in the tubule to any extent. Hence this compound can be used as a marker to indicate the degree of reabsorption of secretion of other substances. If the blood plasma contains a concentration P of inulin and the final urine a concentration of U then the amount of plasma that must have been filtered per minute to give this urine concentration is $\frac{UV}{P}$ ccs/min. This is termed the 'clearance' of inulin. Any substance which has a clearance less than that of inulin must be reabsorbed while a substance with a clearance greater than that of inulin must be secreted into the tubule. This method can be used to study how the kidney treats waste products of metabolism such as urea. If the kidney secreted urea into the tubule its clearance should be greater than that of inulin. This is not found to be the case in man. Urea in the urine of man has a concentration about 70 times that in the plasma as against the 110 times that would be expected even if it were just passively concentrated by the reabsorption of water from the renal tubule. Thus not only is urea not secreted, some is actually reabsorbed. The proportion of the urea reabsorbed is not however related to the concentration in the filtrate as with substances which are actively reabsorbed, so it is presumed that the reabsorption of urea represents a passive leakage from the tubule back into the blood.

Renal regulation of blood reaction

The properties of proteins, the permeability of cell membranes, the active transport of ions and the ion concentrations of cells are among the factors influenced by acidity or alkalinity of the body fluids and hence the pH of the blood has to be finely regulated. The normal pH of mammals is between 7·35 and 7·45 and variations to 7·0 or 8·0 may prove fatal. Three factors act to preserve the normal pH of the blood: (1) buffers in the blood and cells; (2) control over the removal of carbon dioxide by variations in the rate of respiration, and (3) excretion of acid radicals by the kidney. Quantitatively the most important acid produced in the body is carbonic acid of which up to 15,000 mE may be formed daily. The concentration of bicarbonate in the blood is readily controlled by blowing off excess CO_2 in the lungs and qualitatively more important is the small daily excess of non-volatile acids which are taken into the body. These include inorganic acid radicals such as chloride, sulphate and phosphate and organic acids taken in with the food or produced as a result of metabolism. On a normal diet man has a daily excess of such acid radicals of about 40–80 mE over the intake of fixed bases such as sodium, potassium, calcium and magnesium. Excess acid radicals can be accumulated in the blood for a time because being stronger acids than carbonic acid they liberate fixed base from bicarbonates and form carbonic acid. The fixed base neutralises the strong acid and as the carbonic acid is not fully dissociated at the pH of the blood it is less 'acid' than an acid which is fully dissociated. Its formation from bicarbonate therefore buffers the blood against pH change. If the blood becomes more acid relief can be gained by eliminating CO_2 and decreasing the blood carbonic acid. Such relief can be only temporary and it is the responsibility of the kidney to remove non-volatile acids without depleting the fixed base reserve of the body. If the kidney could produce very acid urine or if the urine volume were sufficiently large the excess acid radicals could be removed as strong acid. However, the most acid urine that can be produced by man has a pH of only about 4·7. To eliminate 80 mE of strong acid at this pH would require some 4,000 litres daily! It is clear therefore that mere dilution is not a feasible proposition and that any strongly acidic radicals in the urine must be accompanied by their full equivalent of non-hydrogen base.

Weak acids are only partially ionised in solution and so in

their case a proportion of the acid eliminated need not be associated with fixed base. The pH of the urine and the degree of dissociation of the weak acid are related by the Henderson-Hasselbalch equation.

$$\text{pH} = \text{pK} - \log_{10} \frac{\text{concentration of free acid}}{\text{concentration of salt}}$$

At a pH numerically equal to the pK, $\log_{10} \frac{\text{conc. acid}}{\text{conc. salt}}$ is equal to zero so at this point half the substance must be as free acid and half as a salt. At this pH the system is at its most stable and least sensitive to the addition of free base or acid. If the pH is lower than the pK then more acid will be free and less combined with base, if the pH is higher than pK the greater proportion of acid will be combined with base.

In the urine the most important buffer is the sodium dihydrogen phosphate – disodium hydrogen phosphate system. Bicarbonate is present in higher concentrations in the filtrate than is phosphate but as it is almost completely reabsorbed during the passage of the urine down the tubule it cannot contribute to the buffering capacity when the final urine is acidic. At the pH of the blood about four times as much phosphate is in the disodium hydrogen form than in the sodium dihydrogen form. When the urine is mildly acidic this ratio is reversed and when the urine has its maximum acidity there is about fifty times as much in the sodium dihydrogen form as in the disodium hydrogen state. The fixed base sodium can therefore be conserved at the expense of hydrogen ion. Two main processes seem to be involved. (1) The secretion of hydrogen ion into the renal tubule and the active reabsorption of sodium. The secretion of hydrogen ion involves the formation of carbonic acid inside the tubule cells, the exchange of sodium for hydrogen and the transport of sodium and bicarbonate ions into the blood (Fig. 25).

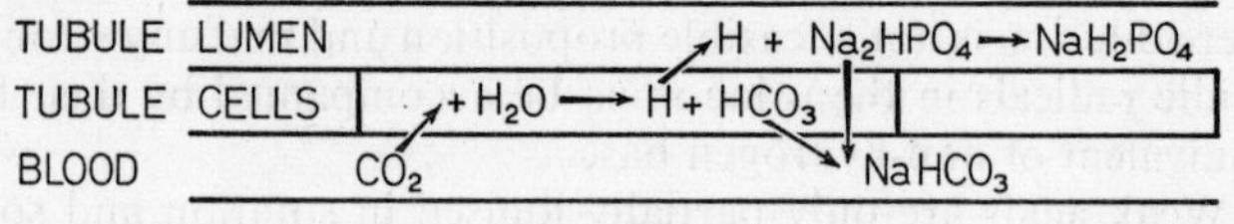

FIG. 25.

Note that if the partial pressure of CO_2 in the blood falls or if the enzyme carbonic anhydrase is poisoned the production of H no longer occurs and the urine becomes alkaline. (2) The re-absorption of sodium and bicarbonate ions from the tubule also helps to cause a shift in the phosphate to the acidic form (Fig. 26).

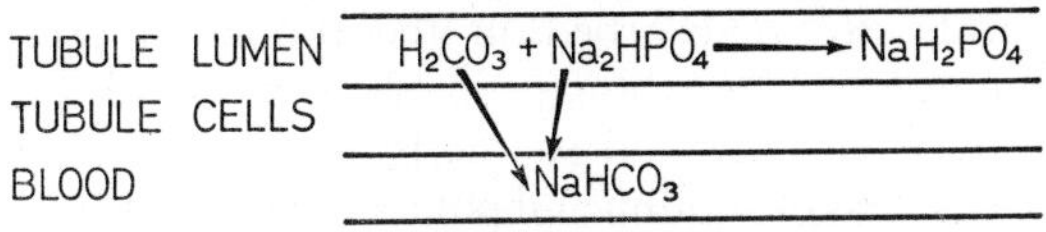

TUBULE LUMEN $H_2CO_3 + Na_2HPO_4 \longrightarrow NaH_2PO_4$
TUBULE CELLS
BLOOD $NaHCO_3$

FIG. 26.

When the urine is very acid the fixed base sodium is still further conserved by exchange for ammonium ions. Ammonia is formed inside the tubule cells from glutamine and alpha-amino acids. It then diffuses down the concentration gradient from cell to tubule lumen and reacts with and neutralises hydrogen ions in the urine. This allows further withdrawal of sodium to occur without the corresponding increase in acidity of the urine which would have accompanied a sodium hydrogen exchange (Fig. 27).

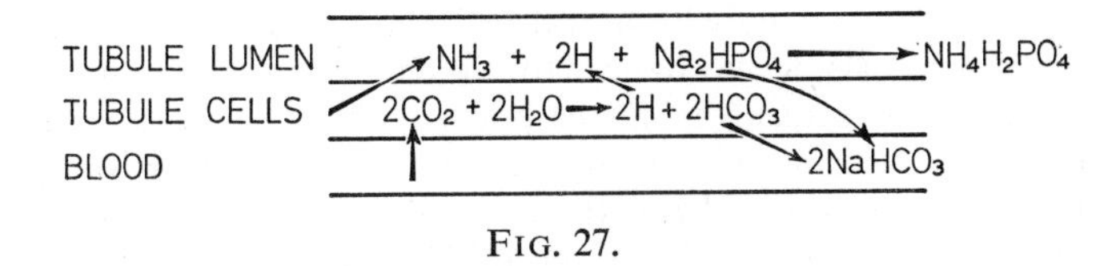

FIG. 27.

About 40 mE of ammonium ion are excreted daily and this amount can be increased tenfold if the blood becomes very acid. All excess acid radicals entering the body can therefore normally be removed as free acid or in association with ammonium ion without involving wastage of fixed base. The success of the kidney in conserving fixed base at the expense of acidic radicals can readily be determined since the fixed base reabsorbed is equivalent to the sum of the ammonium ions in the urine and the difference in the titratable acidity of the urine and the plasma.

It is less common for the alkali production of the body to exceed the acid formation but it can happen that the intake of fixed bases exceeds that of fixed acids. The kidney is less well fitted to produce very alkaline urine and the highest pH is only about pH 8. When alkaline urine is formed fixed acids such as chloride are reabsorbed in the tubule in exchange for bicarbonate. The bicarbonate in the urine is always accompanied by fixed base and so the formation of alkaline urine involves the loss of sodium and potassium from the body. This is rarely important as the alkalosis is usually initiated by an excessive intake of fixed base. In certain conditions involving much sickness the sodium levels of the body may be depleted and then if alkalosis develops there is a conflict between the requirements of ion regulation and that of the reaction. Thus if the fixed base is low and the blood alkaline the kidney should conserve sodium to maintain the body ion levels and excrete sodium to maintain the blood pH. In this case the maintenance of the ion level usually dominates. Similar conflicts between the requirements for the regulation of ion content and body fluid volume are discussed in the next chapter.

In conclusion we may say that the phosphate buffer system in association with the withdrawal of fixed base in exchange for hydrogen ion is the mechanism for the elimination of the normal excess acidic radicals from the body. If there is a large excess of acidic radicals in the body the buffering capacity of the phosphates is strained and the urine pH falls. This evokes the production of ammonia by the tubule cells. Combination of ammonia with hydrogen ion in the urine allows further exchange of sodium for hydrogen to occur and hence a greater elimination of hydrogen relative to fixed base. Bicarbonates are completely reabsorbed when acid urine is being formed. Alkalosis is accompanied by withdrawal of fixed acid and excretion of bicarbonate and fixed base.

The pH of the urine does not change in the proximal tubule so it is probable that the modification of the reaction takes places in the distal tubule.

INVERTEBRATES

The excretory organs of invertebrates may be divided into four main types: (1) nephridia; (2) the 'kidneys' of molluscs; (3) coxal glands; (4) malpighian tubules.

Nephridia

There are three main types of nephridia, flame cells, solenocytes and true nephridia. The first two of these have the proximal end closed and the third usually, though not invariably, opens into the coelom.

Closed nephridia of the flame-cell type are found in the Platyhelminthes and Rotifera. In these excretory organs the proximal intracellular branches of the excretory cells end blindly within the protoplasm of the terminal cells. The lumen of the tubules contain bundles of cilia (the 'flame') whose action serves to agitate the urine and drive it along the tubule. The solenocytes which are found in the Cephalochordate, *Amphioxus*, have the same general structure but differ in detail. It is not known how flame cells and solenocytes form urine but the process probably involves some form of secretion.

The nephridia of annelids and the larvae of molluscs are ectodermal invaginations. The proximal end of the tubule may be ciliated and open to the coelom via a nephrostome or it may be closed. Nephridia occur segmentally in annelids though in some forms secondary multiplication gives rise to many nephridia per segment. Often the nephridial tubules run through the body wall and open independently as is the case in *Lumbricus*, but it is not uncommon for some or all of the nephridia to unite distally and pour their secretion into a common duct which empties into the gut (Fig. 28). This latter situation is found in the common British worm, *Allolobophora*, and in the Indo-Pacific earthworm, *Pheretima*. The excretory tubules of some annelids are compound organs formed by a fusion of mesodermal coelomoduct tubules and the true nephridial tubules. Such a compound nephridia is termed a nephromixium. An example of this condition is found in the lugworm, *Arenicola*.

The mollusc 'kidney'

The molluscan kidney is a modified coelomoduct and is therefore mesodermal. It has a thickened and folded wall and typically opens proximally into the pericardial coelom and distally into the mantle cavity. One pair only is found in most molluscs but two pairs are found in some cephalopods and six

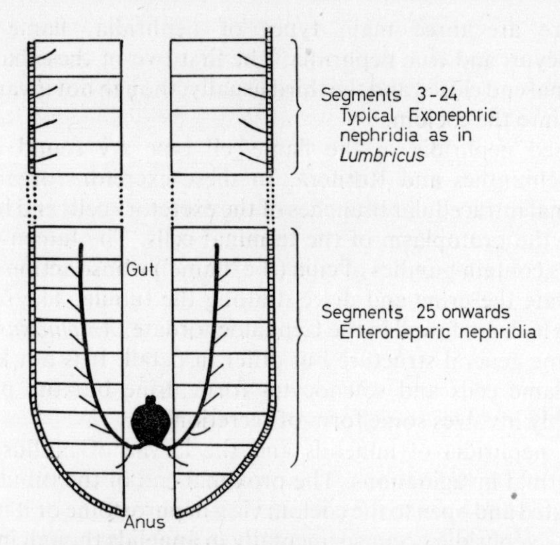

FIG. 28. The exonephric and endonephric nephridia of *Allolobophora*. (Redrawn from two diagrams in Bahl)

pairs in the recently discovered partly segmented 'archimollusc' *Neopilina*. In this last form some of the coelomoducts drain the coelomic cavities that lie above and below the gut and some the pericardial coelom.

Coxal glands

The segmental excretory organs of Arthropods are also coelomoducts. They consist of a closed end sac which represents a coelomic remnant and a mesodermal tubule running to the base of an appendage. In the Onycophora the organs are segmental and in some arachnids such as the king crab, *Limulus*, several pairs may be present. In the crustacea usually only one pair is present opening at the base of the antenna or maxilla. Less frequently both antennary and maxillary glands may be present.

Malpighian tubules

Malpighian tubules in association with glands in the rectum form the excretory complex of most insects, myriapods and terrestrial arachnids. These tubules end blindly and lie in the extracellular space bathed with haemolymph. Distally they open into the posterior part of the gut.

Nitrogenous waste material can readily diffuse across the permeable parts of the body surface in aquatic animals and hence the excretory organs of aquatic invertebrates like those of aquatic vertebrates are primarily concerned with regulation of the ionic composition, concentration, and volume of the body fluids. In terrestrial forms they are also the organs primarily responsible for the removal of nitrogenous waste.

In the annelids and crustacea, the only invertebrates other than the insects whose excretory organ function has received more than cursory study, a correlation can be drawn between the length of the nephridial or coxal gland tubule and the concentration of the urine produced. Marine worms such as *Nereis cultrifera*, the rag worm, have rather short nephridial tubules, *Lumbricus*, which is physiologically a fresh water form, a long tubule. Similarly the lobster, *Homarus*, has a short duct in its coxal gland whilst that of the fresh water crayfish, *Astacus*, is long. *Homarus* produces urine isosmotic with the blood and so probably does *Nereis*, the other two species can conserve ions in the body by producing urine less concentrated than the blood. The correlation between duct length and minimum urine concentration is well shown too by different species of the amphipod, *Gammarus*, from marine, brackish and fresh waters. The marine species, *G. locusta*, has a short duct to its antennary gland, *G. pulex* from fresh water has a long duct and the brackish water species *G. duebeni* has one of intermediate length. *G. locusta* forms urine isosmotic with the blood. *G. pulex* and *G. duebeni* can both form urine hyposmotic to the blood but that of *G. pulex* is more dilute than that of the latter species. It has already been mentioned that the excretory organs function to regulate the ionic composition of the blood even in those forms that produce only isosmotic urine.

Mode of urine formation

The excretory organs of crustacea probably form urine by a filtration process since, as in the case of vertebrates, inulin injected into the blood appears in the urine. The molluscs too are thought to have filtration systems since a back pressure stops urine formation. In advanced terrestrial gastropods such as the giant African land snail, *Achatina*, filtration is from arteries supplying the kidney but the primitive mode of urine formation is by filtration through the wall of the heart into the pericardial cavity. From the latter the filtrate passes through the reno-pericardial canal into the lumen of the kidney.

Many annelids have open nephrostomes to their nephridia and in the earthworms with this type of nephridia it is probable that coelomic fluid may be carried directly into the nephridial lumen since yellow material from the chlorogogenous cells may be found there. It is clear however that nephridia with no internal opening must form urine either by secretion or by filtration from the numerous blood vessels which surround the tubule. Whether formed by filtration or secretion, the urine is modified during passage through the nephridia. Protein is practically absent in the urine though present in considerable quantities in the blood and coelomic fluids. Similarly chloride, phosphates, sodium and potassium are all also present in lower concentrations in the urine than in the body fluids indicating that they are extensively reabsorbed in the tubule. Ramsay has successfully extracted fluid from different regions of the nephridium of *Lumbricus* and determined the osmotic pressure of the samples. The urine is isosmotic with the blood until it enters the middle tube where it is diluted. Dilution is completed in the wide region (Fig. 29).

The coxal glands of crustacea are supplied by a branch of the

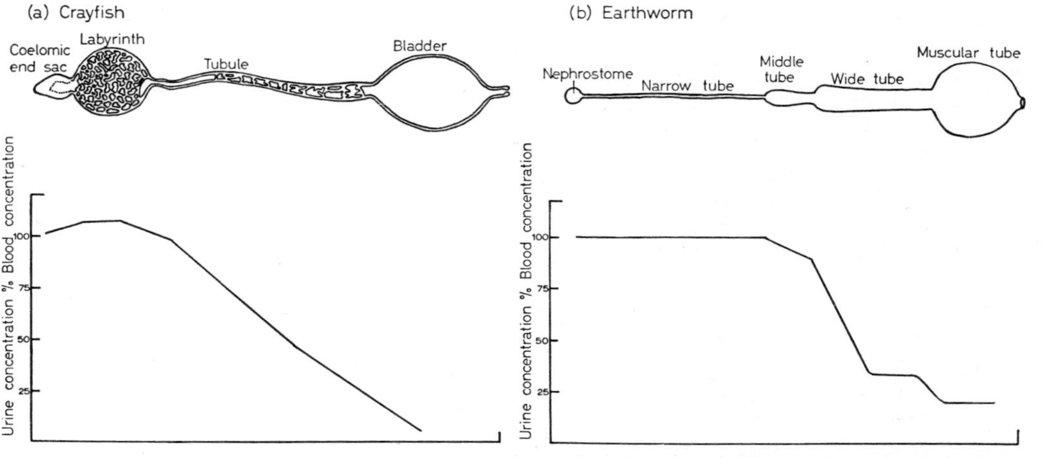

FIG. 29. Semidiagrammatic representation of the parts of the excretory canal responsible for the dilution of the urine in (a) the Crayfish (b) Earthworm. ((a) after Peters (b) after Ramsay)

antennary artery which runs to the coelomic end sac and may also supply the labyrinth if this structure is present. The fluid in the end sac of *Astacus* has been found to have the same concentration as the blood but the concentration falls along the length of the tubule separating the labyrinth from the bladder (Fig. 29). This segment of the tubule is lacking in the lobster. Inulin injected into the blood appears in the urine which lends support to the suggestion that the initial formation of urine is by filtration from the antennary artery even though no obvious site for filtration comparable with the Bowman's capsule of vertebrates is apparent in the end sac.

Malpighian tubules lie freely in the haemocoel of insects and are not supplied by special arterial vessels. Urine formation by ultrafiltration is not therefore possible. The wall of the tubule is, however, fairly permeable to small molecules such as the inorganic ions, amino acids, and sugars and it is thought that urine formation may depend on the secretion of potassium into the tubule lumen from the blood and that water and other small molecules then enter passively. The final composition of the urine is determined by the rectal glands which absorb metabolites, ions and water. The final concentration of the urine may be higher than that of the blood. Insects living in dry environments may reabsorb almost all the water from the faeces and urine and produce a partially dry product.

SUMMARY

Though there is a variety of morphological structures there is a common pattern to the excretion of both vertebrates and invertebrates in so far as the urine is formed initially in the proximal region of the excretory organ and then modified on its way down the tubule. The initial formation of primary urine is either by filtration from the blood or by some form of secretion.

8

The Regulation of Body Water and Ion Content

Vertebrates

Water constitutes some 60–70 per cent of the body weight of vertebrates. The unthinking assumption made by many people that any change in their weight is due solely to variation in body fat content therefore pays a considerable compliment to the effectiveness of the mechanisms responsible for the regulation of body water. There is, however, some justification for this view in as far as daily variations in the water content of man are small, not exceeding about 0·2 per cent of the total in normal individuals. Even the intake of very large volumes of fluid (12–15 litres a day) is so effectively compensated that body weight is still maintained within about 0·5 per cent of the mean. It is important that changes in body water content should be restricted to within such limits. Water is distributed between the intra and extracellular fluids according to their respective solute concentrations. Increase in the concentration of the extracellular fluid draws water from the cells whilst conversely decrease in extracellular fluid concentration results in cellular swelling. Consequently any variation in body water content without adjustments in solute concentration tends to cause shrinkage or swelling of both cellular and extracellular compartments.

Salt loss or uptake also results in a redistribution of water between extracellular and cellular compartments. Loss of sodium chloride from the body dilutes the extracellular fluid and such dilution results in a water shift to the cells as described above. Conversely increase in the salt concentration of the extracellular fluid results in a withdrawal of water from the cells.

In contrast to these effects the uptake or loss of a salt solution isosmotic to the blood has no effect on the concentration or volume of the cells, it affects only the extracellular volume.

The body responds initially to a high salt intake by retaining water and by increasing the intake of fluid. Conversely a low salt diet is initially accompanied by a decrease in body water content. Such an association between salt intake and the amount of water retained in the body would be expected if the body's regulatory mechanisms were directed primarily at the maintenance of the concentration of the blood since water retention would tend to keep the concentration normal when the salt intake was high and vice versa. That the regulatory mechanisms also respond to the volume of the extracellular fluid space is however shown by the fact that the rate at which salt is excreted relative to water is not invariably directed at the maintenance of concentration. Thus if water intake is suppressed over a prolonged period there is an increase in the concentration of the blood. Correction of this would necessitate that the sodium output in the urine should be increased. In fact it decreases. This retention of sodium favours retention of water, and hence volume maintenance, but it naturally aggravates the increase in extracellular ion concentration.

Variations in the rate of salt and water intake result in six qualitatively different conditions to which the regulating mechanisms must respond if the extracellular concentration and volume are to be maintained constant (Table 19).

A glance at Table 19 indicates that in the case of two conditions; (4) where the body water volume is normal but the salt concentration is subnormal, and (5) where the water volume is sub-normal and the salt content normal, the requirements of the mechanisms maintaining concentration and volume are directly opposed. On the other hand in (1) where water is in excess and salt below normal concentration both volume and concentration regulation necessitate the same solution. It is not surprising therefore that the body's response to (1) tends to be more rapid than that to (4) and (5).

TABLE 19.

Total Fluid volume	*Salt in extracellular space*	*Extracellular concentration*	*Correction for volume regulation*	*Correction for concentration*
(1) Excess	Normal	Hyposmotic	Excrete water Retain NaCl	Excrete water Retain NaCl
(2) Normal	Excess	Hyperosmotic	Excrete NaCl Retain water	Excrete NaCl Retain water
(3) Excess	Excess	Normal	Excrete NaCl Excrete water	No correction necessary
(4) Normal	Sub-normal	Hyposmotic	Retain water Retain NaCl	Excrete water Retain NaCl
(5) Sub-normal	Normal	Hyperosmotic	Retain water Retain NaCl	Retain water Excrete NaCl
(6) Sub-normal	Sub-normal	Normal	Retain water Retain NaCl	No correction necessary

Regulation of osmotic concentration of the blood

When a large quantity (1,000 ccs) of water is drunk it is rapidly absorbed from the gut into the blood. After a delay of about 30 minutes in the case of man, the urine output increases and rises rapidly to a maximum some 10 times the normal rate. This high urine output continues until the excess water load has been eliminated from the body. Control over the urine volume in man is achieved almost entirely by varying the amount of water reabsorbed in the distal tubules and collecting ducts, though in many lower vertebrates the glomerular filtration rate may also be changed.

Regulation of the amount of water reabsorbed in the tubules is controlled by the amount of a hormone, the antidiuretic hormone, in the blood reaching the kidney. Antidiuretic hormone (A.D.H.) increases the permeability of the wall of the distal tubule and collecting ducts to water. Hence, when the A.D.H. concentration is high much of the water in the urine which has passed into the distal tubule is withdrawn by osmosis into the interstitial spaces of the medulla as it passes down the collecting ducts. In the absence of A.D.H. the collecting duct walls are less permeable to water, osmotic withdrawal is smaller and the final urine volume is larger.

A.D.H. labelled with radio-active iodine is found to be absorbed on to the wall of the distal tubule and collecting ducts but not on to other regions of the tubule and it is presumed to affect only these two areas.

A.D.H. is released from the neural lobe (pars nervosa, posterior lobe) of the pituitary, a small compound gland suspended from the floor of the hypothalamus of the brain. During the embryological development of vertebrates, an upgrowth from the roof of the buccal cavity joins a downgrowth from the brain to form the pituitary and from this gland originate hormones which directly or via their effect on other endocrine organs affect various physiological functions. A.D.H., though released from the neural lobe of the pituitary does not originate there, but comes from the hypothalamus itself. A.D.H., or its precursor, migrates down the nerve tracts joining the hypothalamus to the pituitary. If lesions are made in the hypothalamus or alternatively if the nerve tracts of the pituitary stalk are severed, A.D.H. disappears from the neural lobe and, though it may still be present in the hypothalamus proximal to the block, is no longer released into the blood stream in an effective form. In experimental animals lesions of this type result in marked increases in urine volume. Naturally occurring lesions of this type also occur in man and give rise to *diabetes insipidus*, a condition similarly characterised by excessive urine production.

The rate at which A.D.H. is released from the pituitary is dependent on the osmotic concentration of the blood bathing the brain. The injection of small quantities of NaCl, Na_2SO_4, or sucrose hyperosmotic to the blood into the internal carotid artery of dogs is followed by a transient but marked decrease in urine production. The extent of the response is related to the concentration of the solute injected. Isosmotic volumes of sucrose are almost as effective as NaCl in stimulating anti-diuresis which suggests that the receptors responsible for monitoring the blood concentration (osmo-receptors) respond to the overall osmotic concentration and not to the concentration of the inorganic ions of the blood. Injection of solutions isosmotic with the blood does not affect the urine volume. Not all substances are equally effective in promoting anti-diuresis. Glucose produces some, though of a lesser extent than equivalent concentrations of NaCl, but injections of hyperosmotic urea have no effect on the urine volume. The interpretation put on these observations is that initiation of A.D.H. release is dependent on volume changes of certain specialised osmo-receptor cells. The cell wall of the osmo-receptors is assumed to be impermeable to NaCl, Na_2SO_4, and sucrose, slightly permeable to glucose, and more freely permeable to urea. Hence, hypertonic solutions of NaCl, Na_2SO_4, and sucrose would be expected to withdraw water from the receptors by osmosis; but urea, since it can penetrate the cell wall would have no such effect.

It is possible that the failure of the osmo-receptors to respond to increased concentrations of urea may have been of some importance to the elimination of this waste product by early terrestrial vertebrates. Urea removal is dependent on urine volume in some lower vertebrates such as *Rana* and lung fish. Hence, decrease in urine volume would tend to diminish the rate of removal of urea from the blood. If the osmo-receptors responded to raised urea levels in the blood by decreasing urine production a vicious circle would be set up whereby urea retention raised the blood concentration and by bringing about a decrease in urine volume further raised the blood concentration.

The site of the osmo-receptors has not been positively recognised but it may be associated with vesicles lying in the supra optic nucleus. In the dog the total surface area of the vesicles is less than one square mm and if these are indeed the osmo-receptors it is interesting that the integrity of the whole body is ultimately dependent on the permeability of so small an area.

Although the concentration of the blood is the main factor regulating the release of A.D.H., some regulatory effect is exercised by other brain centres. Emotional conditions such as fright, anger or pain all cause an increase in A.D.H. release. Thus a few minutes after a dog is subjected to stress, the pituitary releases up to ten times the amount of A.D.H. required to give a maximum antidiuresis. (This should of course be distinguished from the stimulus to empty the bladder which also accompanies stress.)

Antidiuretic hormones and their functions

Extracts of the neural lobe of the pituitary contain not one but two hormones which, in man, are oxytocin and arginine vasopressin. Both are octopeptides, but they differ in two amino-acid groups. Until recently only rather crude extracts of the posterior pituitary were available and hence the specific functions of the two hormones have not been extensively studied. Now that these hormones have been artificially synthesised, however, it has been found that in man oxytocin has little effect on either water or salt output, vasopressin being primarily responsible for the antidiuretic effects considered above. In contrast in mammals such as the rat which have more labile filtration rates than man, oxytocin may modify the renal blood supply and hence affect the glomerular filtration rate. Yet another effect is found in the dog where both oxytocin and vasopressin affect the glomerular filtration. In vertebrates other than mammals arginine vasopressin is replaced by arginine vasotocin from which it differs in one amino acid substitution.

Extracts of the posterior pituitary also bring about anti-diuresis in birds, reptiles and amphibians but not apparently in fish. The degree of response in amphibians is correlated with their mode of life. Terrestrial forms are more likely to be exposed to water shortage than aquatic types and tend to respond more effectively to antidiuretic hormone injections. Thus anti-diuresis is produced in the toad *Bufo* by a dose of pituitary extract only 1,000th of that required to produce a similar anti-diuresis in the fully aquatic toad *Xenopus laevis*. The greater sensitivity of the terrestrial form clearly reflects the difference in the problem of the maintenance of water balance in terrestrial and aquatic forms, *Xenopus* usually having to 'bail out' excess water taken up by osmosis, whilst *Bufo* must restrict water loss when in dry surroundings. Both reduction in filtration rate and increased tubular reabsorption play a part in the antidiuresis of amphibians.

In amphibia the action of posterior pituitary hormones is not limited to the kidney. In mammals as we have seen the action of vasopressin is to promote increased permeability of the walls of the distal tubule and collecting ducts. Similarly injections of posterior pituitary extracts bring about an increase in the water permeability of frog skin and hence increase the osmotic uptake of water. Water withdrawal from the bladder is increased and so also is the rate of sodium uptake by the skin. It will be noted that all these effects tend to conserve or increase the water content in the body.

Salt regulation

Salt regulation is largely independent of the water balance hormones. Dogs with *diabetes insipidus* produced by surgical severance of the hypothalamic-hypophyseal tract eliminate salt absorbed from the gut in about the same time whether or not they are given injections of posterior pituitary to control the urine volume. The total salt output by these dogs is thus fairly constant even though the concentration of salt (Cl) in the urine varies inversely with the urine volume.

Regulation of the rate of salt loss from the body is governed by hormones released from the adrenal glands, a pair of endocrine organs lying in close proximity to the kidneys. The adrenal of mammals is a compound structure. The inner or medullary region is innervated by the sympathetic system and on stimulation releases adrenalin and noradrenalin. The outer or cortical region, which is not innervated, produces hormones which regulate salt and sugar metabolism. All the active principles which have been identified so far from this region are steroids.

The basic structure of a steroid is shown in Table 20 together with the structure of the hormones believed to have the most potent effect on metabolism. Many other less active principles have been extracted from the adrenal cortex. The non-sexual hormones have sometimes been divided into two groups; those that exert their influence on general metabolism by bringing about a rise in the blood sugar level and increasing fat and protein metabolism (glucocorticoids) and those primarily concerned in mineral metabolism (mineralocorticoids). The distinction between mineralocorticoids and glucocorticoids is, however, best to be regarded as one of degree not of kind as any effect on sugar metabolism affects salt metabolism to some extent and vice versa. Thus cortisol, which is primarily a glucocorticoid, increases the retention of sodium and loss of potassium by the kidney as does the mineralocorticoid, aldosterone.

TABLE 20.

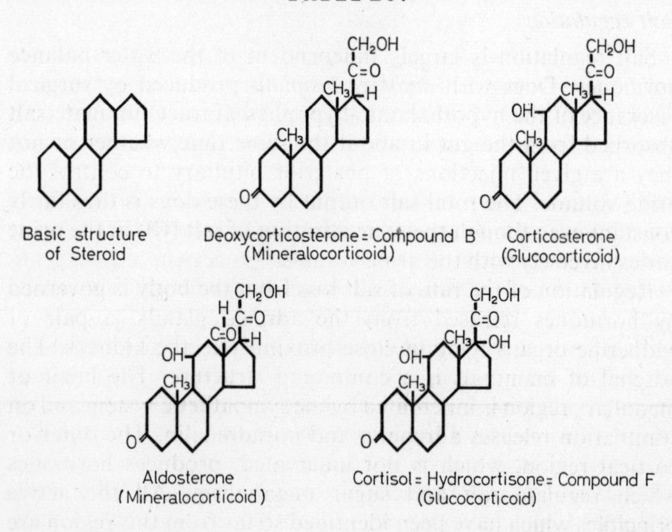

One of the most important actions of the adrenal corticoids is to stimulate the reabsorption of sodium by the renal tubules, and hence determine the amount of sodium retained in the body. Complete absence of aldosterone results in a rise in the sodium loss in the urine to about 10 per cent of that filtered initially instead of the normal 1–3 per cent. Corticoids also regulate the relative proportions of ions excreted in the urine. Under the influence of aldosterone and deoxycorticosterone at least part of the sodium withdrawal from the renal tubule is in exchange for potassium or hydrogen ions. As most, if not all, the potassium entering the proximal tubule with the filtered plasma is reabsorbed before the primary urine reaches the distal tubule this exchange of potassium for sodium must be of considerable importance in regulating potassium excretion.

The action of mineralocorticoids, like that of A.D.H., is not limited to the kidney. When the colon of the dog is treated with deoxycorticosterone the movement of sodium from lumen to blood is increased but the movement in the reverse direction remains unaffected. Consequently the net uptake of sodium from the gut is increased. At the same time the net loss of potassium to the gut lumen is also raised. In sweat glands sodium is believed to be secreted by the proximal region and reabsorbed distally as the secretion is diluted. Corticoids affect the concentration of both sweat and saliva. When deoxycorticosterone is injected into man there is an excessive reabsorption of sodium and the saliva and sweat are abnormally dilute. The action of corticoids is thus aimed fairly generally at sites where there is sodium reabsorption. They do not affect areas where there is not normally uptake of sodium. Thus deoxycorticosterone apparently has no affect on the small intestine.

The action of the mineralocorticoids in restricting salt loss in the kidney, colon, sweat and salivary glands indirectly results in an increase in blood volume as water is retained in the interest of the maintenance of blood concentration.

Mineralocorticoids, or at least aldosterone, may also play a more direct part in the retention of water in the body. When over-hydrated human subjects are given injections of A.D.H. the urine volume is decreased and the concentration of the urine is raised. However, if the dose of A.D.H. is accompanied by aldosterone the urine volume is still further decreased and the concentration raised even higher (Table 21). The interpretation drawn from these results is that aldosterone, by stimulating sodium reabsorption in the loop of Henle, further raises the concentration in the renal medulla and hence increases the

TABLE 21.

	ADH alone	*ADH plus Aldosterone*
Urine volume	413 ccs/8 hours	326 ccs/8 hours
Urine conc.	955 m. osmoles.	1037 m. osmoles.

Data from Crabbe, J.

osmotic withdrawal of water from the collecting ducts. The final urine volume is decreased and the concentration increased. The action of A.D.H. is thus supplemented.

These results may also be used to draw attention to an important point in the interpretation of measurements of urine concentration. This is that the overall effect of urine production on the salt balance of the body can only be determined by considering both the urine concentration and the urine volume and not just the former. In the experiment cited above the urine becomes more concentrated under the influence of aldosterone and at a casual glance it might therefore be thought that aldosterone was increasing the sodium loss from the body. A simple calculation shows, however, that the reverse is the case, the total m. osmoles lost in the eight hour period being almost 17 per cent higher when the aldosterone was absent.

The importance of the mineralocorticoids in regulation of the body fluids is made abundantly clear if the adrenal cortex is removed. Cessation of corticoid production is followed by excessive loss of sodium chloride from the body and by retention of potassium. Associated with the salt loss there is a marked decrease in extracellular volume and blood pressure. If the subject remains untreated, death occurs after about two weeks.

Control of adrenal hormone production

The adrenal cortex is not innervated and as it does not apparently respond directly to variations in the corticoid content of the blood bathing the organ it is clear that the regulation of the rate of hormone release must itself be under hormonal control. Nevertheless, the response of the adrenal cortex to conditions requiring increased hormone release can be quite rapid. For instance when the blood volume of dogs is suddenly decreased by bleeding the aldosterone concentration in the blood begins to rise within a few minutes and may be two to five times the initial value after about twenty minutes.

The liver is the main site of the deactivation of aldosterone circulating in the blood so any change in the rate at which breakdown occurs will also have an indirect effect on the salt balance of the body.

The means by which the adrenal cortex secretion is regulated is far from fully worked out but it seems likely that at least three hormones may be involved in controlling the output of the mineralocorticoids and glucocorticoids. One of these hormones, Adrenocorticotrophic hormone (Corticotrophin, A.C.T.H.) has been identified but hormones from the kidney and pineal region of the brain diencephalon are as yet uncharacterised.

Adrenocorticotrophic hormone is a large polypeptide, though it is possible that the hormone is naturally released in the form of a protein. It is formed and released by the anterior lobe of the pituitary. The anterior pituitary can be stimulated by the level of circulating corticoids in the blood reaching the hypothalamus, increasing A.C.T.H. output when the corticoid level is low and vice versa. A.C.T.H. is also released in large amounts when the animal is exposed to a variety of conditions producing stress. If the pituitary is isolated from the brain it will still continue to release A.C.T.H. when stimulated with substances such as histamine, adrenalin and 5 – hydroxytryptamine but it no longer responds when the animal is stressed. It may be presumed therefore that the response to stress is regulated by brain centres. The nerve supply to the anterior pituitary is small and does not appear to be implicated in the control of A.C.T.H. Secretion is mediated by yet another hormone, the corticotrophic releasing factor (C.R.F.) At one time it was thought that C.R.F. and

A.D.H. (vasopressin) were one and the same as not only do injections of vasopressin stimulate A.C.T.H. production but also A.C.T.H. is produced naturally in several conditions which stimulate A.D.H. release. However, when the blood is diluted and its volume increased by the intake of a considerable amount of water, A.D.H. release is suppressed but A.C.T.H. is released. Furthermore, the dose of vasopressin required to stimulate A.C.T.H. release is about 1,000 times that required to give a maximal diuresis. A.D.H. cannot therefore be the same as C.R.F. It is probable however that C.R.F., like A.D.H., is a polypeptide. The blood supply to the anterior pituitary comes via a portal system from the hypothalamus and it appears that C.R.F. passes in this portal system from the mid hypothalamus, as blood taken from this portal system in stressed dogs stimulates A.C.T.H. production in rats.

The function of A.C.T.H.

When A.C.T.H. output is increased the rate of corticoid release from the adrenal glands is raised considerably. However, though all corticoids are affected by A.C.T.H. it appears that the target of this hormone is the glucocorticoids as their level in the blood is influenced more than is that of the mineralocorticoids. A.C.T.H. is therefore primarily concerned with the release of hormones which stimulate general metabolism not mineral metabolism. This is further confirmed by the observation that removal of the pituitary has a more severe effect on general than on mineral metabolism. Removal of the pituitary has been found to reduce the output of glucocorticoids to about 10 per cent of the normal value while aldosterone output falls by only a little over 50 per cent.

Regulation of the output of aldosterone

The mechanisms regulating aldosterone production have only come under extensive study in the last few years and the whole subject is still highly controversial. In the interests of providing a simplified account the information presented in this sub-section is biased in that it sets out the views of only some of the protagonists. The reader is therefore warned against a too facile acceptance of the evidence presented and urged to consult the more detailed evidence set out in the papers listed for further reading.

Aldosterone output from the adrenal glands can be varied in animals lacking a pituitary gland and hence release of this hormone must be influenced by factors other than A.C.T.H. Nevertheless brain stem lesions are found to affect the rate at which aldosterone is released indicating that as in the case of A.D.H. and A.C.T.H. the central nervous system is implicated in the control of salt balance. The active site in the brain is thought to be localised in the diencephalon in the region of the pineal stalk. Injections of extracts of the diencephalon of cattle bring about an increase in the rate of hormone release from the adrenals.

The pineal gland is indicated as a centre involved in salt metabolism as its histological appearance is found to change when rats are kept on a low salt diet. The chemical nature of the hormone, the glomerulotrophic hormone, stimulating aldosterone production is as yet unknown.* Aldosterone output is affected both by the sodium concentration and by the volume of the blood. If the blood concentration is low but the volume normal, aldosterone output is increased. This effect may, however, be antagonised if the blood volume is above normal or may be amplified if the volume is below normal. It is not yet known where the receptors responsible for the response to concentration are located, but those detecting changes in the blood volume are, as might be expected, in the central parts of the blood system itself. Expansion of the right auricle or even pulling on its wall has the effect of bringing about a decrease in aldosterone production and a similar result follows increase in the pulse pressure of the carotid artery. Conversely, a decrease in the blood supply to the great veins produced by constriction of the anterior vena cava or decrease in the carotid pulse, increases the production of aldosterone. The vagus nerve apparently carries

* Recent studies suggest that it may be 6 – methoxy – 1 methyl – 1, 2, 3, 4 – tetrahydro – 2 – carboline.

to the brain the stimuli which result in the suppression of aldosterone production, as severing it prevents the response. It is not, however, involved in conveying stimuli which increase the output of the hormone. Recent reports suggest that the kidneys may also produce a hormone which stimulates aldosterone production. When the kidney blood supply is diminished, as by excessive bleeding of the animal, aldosterone output by headless animals is increased.

Regulation of blood volume

From the evidence presented in the foregoing paragraphs we can erect a hypothesis suggesting the means employed by the body in the regulation of the extracellular volume. If the blood volume is below normal, aldosterone production is reflexly increased by a chain of reactions involving receptors in the carotid artery and great veins, which eventually stimulate the release of the hormone glomerulotrophin.

Increase in aldosterone concentration in the blood stimulates sodium retention by the kidneys and other sites thus raising the concentration of the blood. This rise in the concentration of the blood in turn acts on the hypothalamic centres responsible for the regulation of A.D.H. release and brings about antidiuresis. If water intake is possible the blood volume will be raised. A contrary chain of events will return a raised blood volume to normal.

There is also some evidence that this chain of events may be expedited by a more direct action of changes in blood volume on A.D.H. output. Thus when the left auricular wall is stretched it is found that the output of A.D.H. is diminished. Changes in urine volume associated with different body postures may also be associated with the reflex control of A.D.H. output. A change from a sitting to a standing position, which tends to result in accumulation of blood in the legs and restricts the circulatory volume, results in a temporary decrease in urine output. Adoption of a prone position increases urine output; but sleep induces antidiuresis. Other factors which influence circulating volume also affect urine volume. Thus exposure to cold, which results in transference of fluid from the superficial capillary beds to the main circulating pathways, is associated with increased urinary output.

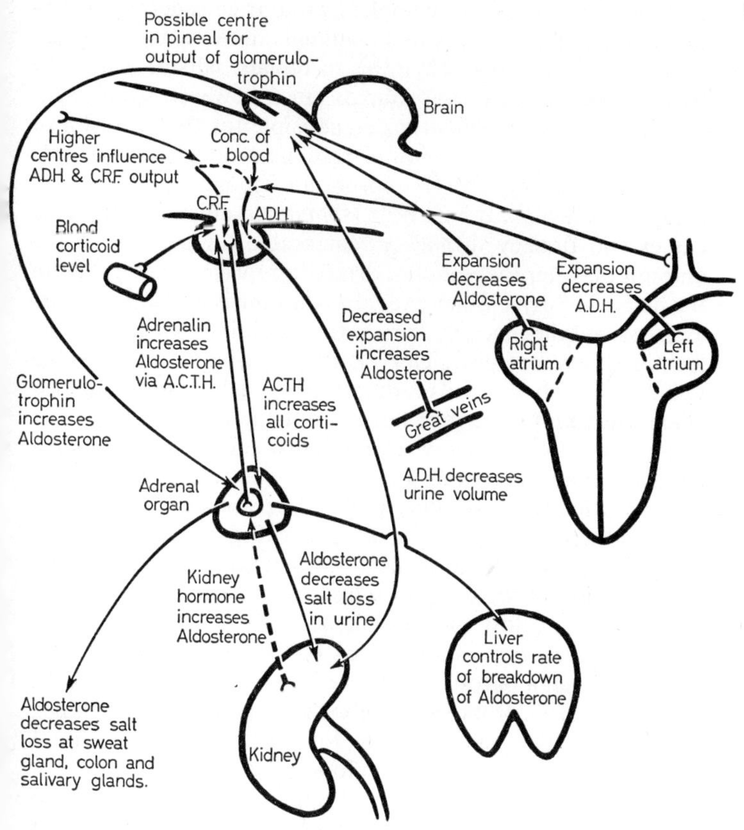

FIG. 30. The sensory and hormonal systems involved in the regulation of the water and salt balance of mammals. (See text for explanation.)

The effects of various conditions on the mechanisms responsible for regulation of blood volume of vertebrates are summarised in Fig. 30.

Control of salt and water intake

An account of the mechanisms involved in body fluid regulation would not be complete without considering the uptake of salt and water. Aquatic animals need recourse only to purely physiological mechanisms to replace salts or water lost from the body. Terrestrial animals must, however, supplement the physiological mechanisms responsible for the retention of ions and water by behavioural responses designed to enable them seek out and take in appropriate amounts of these substances. It has already been noted that desert animals such as the camel and donkey display a remarkable 'awareness' of the amount of water that must be drunk to replace water lost from the body. The drive which precedes drinking is well known to us as thirst, but the precise nature of the conditions which make us feel thirsty are less clear. Thirst is stimulated not only when the volume of the body fluids is decreased by dehydration but also when the volume is above normal but the concentration also raised. Man feels thirsty when either his body weight has fallen by about 0·5–1 per cent as a result of water loss or if the blood concentration is raised sufficiently to decrease the water content of cells by about 1–2 per cent. Thus drinking sea water does not slake thirst as it raises the concentration of the blood.

In addition to the drive to replace lost water there is also a drive to replace lost salt. Sheep, which have previously been subjected to excessive salt loss by collecting their saliva continuously from a parotid fistula, will replace the salt loss when allowed to drink from salt solutions. That man has a similar drive to replace lost salt has long been recognised (and capitalised on in human communities by taxes on salt).

Grasslands within a few tens of miles from the sea receive a certain amount of salt from wind-blown spray, but inland continental regions may have pastures containing very little sodium or chloride. The salt available for herbivores in these areas is therefore restricted and many grazing animals supplement their salt intake by making periodic migrations to regions, the so-called bracks or salt licks, where the soil has a high salt content. Journeys by antelopes to salt licks have been noted in Eastern and Southern Africa and even elephants have been known to dig pits in salty soil. Of the central mechanisms which initiate such salt drives nothing is yet known.

Volume regulation of invertebrates

The rate of uptake and loss of salts and water by many aquatic invertebrates is quite rapid. The crayfish, *Astacus*, loses about 8 per cent of its body weight per day as urine whilst the water flea, *Daphnia*, has been calculated to produce a volume of urine equivalent to twice its body water content in 24 hours. Salt losses too may be rapid in invertebrates. The fresh water isopod *Asellus* for instance loses about 1 per cent of its sodium per hour. We may expect therefore that the regulation of body volume must present similar problems to invertebrates as to vertebrates.

The study of the mechanisms responsible for control of the body volume of invertebrates has not yet reached an advanced state but as in vertebrates there is some evidence which implicates the central nervous system in the regulation of at least insects, molluscs and crustacea.

When the sub-imaginal nymph of the bug *Rhodnius* feeds it may take in blood equivalent to ten or twelve times its own weight! Most of the water and salts in this food are absorbed into the animal's blood and then rapidly eliminated via the malpighian tubules. Maddrell has found that a hormone released from the central nervous system is responsible for increasing the urine output after the blood meal. Isolated malpighian tubules placed in a drop of *Rhodnius* blood secrete urine for a time and then stop. If fresh blood from a fed insect is then added secretion starts once more. Blood from an unfed insect does not activate the tubules. The hormone is released from neurosecretory cells in the mesothoracic ganglionic mass. Severance of the connectives to this ganglion interferes with the production of hormone after a blood meal and this suggests that the neurosecretory cells are themselves stimulated indirectly.

The eyestalk of Decapod crustacea contains a large nerve mass and a hormone can be extracted from this region which disturbs the water balance of vertebrates. The removal of the eyestalks of a crayfish does not interfere with the capacity of the animal to increase the rate of sodium uptake but does cause disturbance in the water uptake at the next moult.

The pleural ganglion of snails is thought to be associated with the control of body water content. When these ganglia are removed the water content of fresh water snails rises.

Blow-flies, *Phormia*, appear to have a sense equivalent to that of thirst in man since they drink when short of water. Flies that are drinking, stop when injected with a large volume of distilled water or hyperosmotic saline but flies that are not drinking do not start to drink after injection of a volume of hyperosmotic saline. This evidence supports the conclusion that volume rather than concentration is the important underlying effect behind the stimulus to drink.

The stimuli which regulate drinking appear to be carried in the recurrent nerve since when this is cut the flies drink until bloated.

Regulation of the volume of interstitial fluid in vertebrates

The distribution of fluid between the blood system and interstitial spaces is governed by the hydrostatic and colloid osmotic pressure of the blood. At the arteriolar end of capillaries the hydrostatic pressure of the blood exceeds the colloid osmotic pressure. The capillary wall is readily permeable to small molecules and hence an ultrafiltrate of plasma is forced into the interstitial space. Hydrostatic pressure falls along the length of the capillary and colloid osmotic pressure rises as the plasma proteins are concentrated by the movement of saline to the interstitial space. Eventually the colloid osmotic pressure exceeds the hydrostatic pressure and fluid begins to move back into the capillary. In a normal man a volume of fluid approximately equivalent to the whole of the blood volume is extruded from the capillaries *every minute*. Most of this is reabsorbed once more. The excess, only about 2,000 to 5,000 ccs per day, is ultimately conveyed to the lymph ducts and discharged back into the blood. Despite variations in the rate of lymph production the volume of the interstitial space varies widely if the blood protein concentration or hydrostatic pressure is varied. Increase in arterial pressure or venous back pressure and decrease in blood colloid O.P. all result in a net shift of fluid from the blood to interstitial spaces. An erect posture tends to increase the venous pressure in the limbs sufficiently to cause intersititial fluid to accumulate there. If there is no movement of the limbs to assist venous return to the heart, accumulation of fluid may be sufficiently severe to decrease the circulatory volume and blood supply to the brain. This is one of the reasons for the occasional collapse of soldiers on parade.

Decrease in arterial pressure or an increase in the colloid O.P. of the blood depletes the interstitial fluid and increases blood volume. Control over the balance between blood and interstitial volumes is exercised by the central nervous system by means of adjustments of the peripheral resistance. Any drastic change in the protein concentration however results in shifts of fluid between the two compartments. These shifts help to ensure that the blood circulation will not be rendered sluggish if the total extracellular space is larger than normal or fail if the extracellular space is too small. A large volume of isotonic saline absorbed from the gut is not rapidly excreted and consequently tends to enlarge the blood volume and dilute the blood proteins. This dilution and the associated rise in blood pressure ensures that much of the excess fluid is transferred to the interstitial space. Conversely loss of salts and water from the body concentrates the blood proteins and again the circulatory volume is maintained at the expense of the interstitial space. Dehydration of the body results in an equal concentration of all body fluids but the volume of the blood is again less affected than that of the interstitial space.

Excessive hydration or decrease in the volume of the interstitial fluids interferes with the exchange of oxygen and metabolites between cells and blood but, as is indicated in the following table, the tolerance of volume change is considerably larger than that of the other two main compartments.

TABLE 22.

		Normal Volume, litres	*Approximate percentage change tolerated*
Intracellular		30	–25 to +25
Interstitial	extracellular	12	–40 to +250
Blood plasma	extracellular	3	–50 to +85

Modified from Reeve, Allen and Roberts.

We have noted that the volume of the extracellular fluid is associated with the amount of salt present in the body. The elimination of excess salt is unlikely to have been a problem to the early terrestrial forms living near fresh water, consequently the vertebrate kidney and the mechanisms associated with its regulation tend to be geared to retain salt in the body. Elimination of salt therefore tends to be slow. However, when there is a large salt intake over a long period, the body fluid volume does not increase indefinitely. Some Japanese who feed largely on salted fish have a daily salt intake of about 20 to 30 grams, twice the European level. These people tend to have very high blood pressure. Doubtless the high blood pressure increases the glomerular filtration rate thus compensating for the tendency to retain fluid in the body. The price paid by the Japanese for the maintenance of an excessive blood pressure is a high death rate from cerebral haemorrhage.

SUMMARY

Body water content varies very little from day to day in man. Regulation depends on mechanisms directed at (a) the maintenance of the concentration of the blood, and (b) the maintenance of blood volume. Either mechanism may become dominant according to the needs of the body. For example, salt may be retained in the body if the volume is too low even though the blood be too concentrated. Water is eliminated if the blood is too dilute.

The rate of water removal from the body is regulated by the antidiuretic hormone A.D.H. which acts on the distal tubule and collecting ducts of the kidney. A.D.H. is produced in the hypothalamus and released from the pituitary gland. Its release is controlled by the concentration of the blood passing through the hypothalamus and also possibly by the degree of expansion of the left auricle of the heart. Suppression of A.D.H. release increases urine production. Much A.D.H. in the blood decreases urine volume.

The salt content of the body is regulated primarily by the steroid Aldosterone which is produced in the adrenal cortex. This hormone promotes the retention of sodium in the body by causing increased reabsorption of sodium by the kidney, sweat glands, salivary glands and colon. The rate of secretion of aldosterone is governed by other hormones. A.C.T.H. is less effective in causing aldosterone release than that of glucocorticoids and hormones possibly from the pineal region of the brain and from the kidney seem to be the major factors regulating its output. The release of the hormone from the brain is claimed to be influenced by the degree of expansion of the great veins, right auricle and carotid artery. Aldosterone is metabolised by the liver which thus also plays a role in regulating the level circulating in the blood.

Little is known of the control of water balance in invertebrates but the central nervous system again seems to be implicated.

Appendix 1

Some functions of inorganic ions

The functions of inorganic ions in biological systems are many and diverse. Some of the more important include (1) the activation of enzyme systems; (2) the stabilisation of proteins in solution; (3) the development of electrical excitability; (4) the regulation of the permeability of membranes and (5) the maintenance of a dynamic state of isotonicity between cells and the extracellular fluid.

The properties of ions depend essentially on their valency and atomic number and hence on their tendency to form complexes with water and organic molecules. Proteins being large molecules have potentially many free positive and negative groups and these may be associated by electrostatic interaction either with water or with ions of opposite charge. The presence of salts in solution modifies these electrostatic associations. Consider, for example, the effect of salts on albumins and globulins. In the case of albumins, the tendency for the free charged groups to form associations with water is greater than the tendency for interaction with groups of opposite charge on neighbouring molecules. Albumins are therefore soluble in distilled water. Globulins on the other hand show a greater interaction between neighbouring molecules than with water and consequently this class of proteins is insoluble in pure water. The addition of a quantity of inorganic salts decreases the protein to protein interaction and hence not only renders globulins soluble but also increases the range of temperature and pH which albumins withstand before being precipitated.

Salts in very high concentrations absorb much of the free water in a solution and hence tend to precipitate proteins. In this action the valency of the ions involved is important. The higher the valency the greater the efficiency an ion has in precipitating proteins—hence the use of trivalent ions in the styptic pencils used to stop the bleeding from shaving cuts.

In many if not in all cases enzymes depend on the three dimensional configuration of the protein molecule in order to be able to react with their substrate. Modification of the structure of an enzyme affects its properties. This is one of the main reasons why cells are so sensitive to changes in pH, ion concentrations, and high temperatures, all of which are likely to alter the stereochemistry of enzyme molecules.

In addition to generalised effects on stereochemistry, certain ions seem to have more specific functions in the activation of enzyme systems. Various types of function are possible. An ion which activates an enzyme may (a) form an integral part of the enzyme molecule; (b) serve to link enzyme and substrate; (c) cause a shift in the equilibrium position of the reaction, or (d) act indirectly by releasing ions which inactivate the enzyme system.

The following account does less than justice to the vast literature on the functions of ions.

Sodium

In almost all animals sodium is the main cation of the extracellular fluids. It therefore accounts for the major part of the cation osmotic pressure of the blood and interstitial fluid. Change in the permeability of the cell membranes of excitable cells is responsible for the development of action potentials.

High concentrations of sodium inside cells are deleterious as sodium inhibits some enzyme systems, particularly those associated with glycolysis, and is less active than potassium in activating others. As a monovalent ion sodium tends to offset the action of small amounts of divalent ions in decreasing the permeability of cell membranes.

Potassium

Potassium is the major cation of cells. In addition to the part it plays in the establishment of the membrane potential of cells it activates certain enzyme systems such as pyruvic phospherase and fructokinase. An adequate concentration of potassium must be present in the extracellular fluid if sodium extrusion from cells is to occur normally. In cells the potassium is not uniformly distributed, its concentration in mitochondria being higher than in the general cell sap.

Calcium and Magnesium

Perhaps the most important function of calcium is that it decreases the permeability of cell membranes to water and ions. This effect is of especial importance in the case of excitable tissues. Muscles in calcium free media initially display spontaneous activity, later they lose their excitability. Isolated nerves and muscles swell in salines lacking calcium, probably because the mechanism responsible for the extrusion of sodium can no longer match the faster rate of sodium entry through the more permeable cell membrane. Calcium is also associated with the processes involved in the shortening of the contractile elements of muscles and part at least of this effect is due to the activation of myosin A.T.P. – ase, the enzyme reaction which provides the energy for contraction. Calcium also probably plays some part in the development of action potentials as repetitive stimulation of nerves increases their calcium concentration.

High concentrations of calcium are deleterious to cells as some enzyme systems are inhibited by this ion. Thus it antagonises the activation of pyrophosphatases by potassium.

As a divalent ion calcium is important in stabilising colloids particularly the intercellular cement which binds cells together. In this function magnesium behaves similarly though usually less effectively. However the intercellular matrix of the tissues of *Mytilus* is more effectively stabilised by magnesium than by calcium. In the absence of calcium and magnesium cells tend to separate.

Calcium increases the release of the transmitting agent at the neuro-muscular junction of vertebrates and this action is antagonised by magnesium which inhibits the release of acetyl choline.

When it is present in high concentrations, magnesium inhibits the neuro-muscular junction unless sufficient calcium is present to neutralise its effect. Magnesium is present in the cells of terrestrial vertebrates at a concentration of some 50 times that in the blood. It is not uniformly distributed in cells having a higher concentration in the mitochondria and nuclei than in the sap. High concentrations of both magnesium and calcium depress the oxygen consumption of cells, and possibly this effect may be linked to their action in decreasing the permeability of membranes as the effect is offset by increases in the concentration of potassium. Magnesium is an essential activator of many of the enzymes involved in energy transfer, hence its presence in the mitochondria. Among these can be included A.T.P.–ase, pyruvic phosphatase and fructokinase.

Hydrogen ion

The chemical properties of proteins change with pH since they are ampholytes behaving as acids on the alkaline side of their isoelectric point and as acids on the acidic side. The hydration of proteins is governed by the pH, the water absorbed being at a minimum at the isoelectric point. However, as already mentioned, the presence of other ions may modify the degree of hydration. Enzymic activity is usually at its maximum close to but not always at the isoelectric point. Thus pH change may effect a variety of factors such as colloid osmotic pressure, inhibition of water by gels and enzyme activity. An increase in the acidity of the blood is followed by a loss of potassium from cells as a result of an exchange of hydrogen for potassium ions and an inhibition of the sodium pump.

Anions

Rather less is known of the functions of inorganic anions apart from the buffering actions of phosphates and bicarbonates in cells and in the blood.

High concentrations of phosphates tend to inhibit calcium actions possibly by lowering the solubility of the calcium. Thus cells are more likely to separate in low calcium media if high concentrations of phosphate are present.

Bicarbonate stimulates the respiration of isolated tissues and the presence of this ion in the bathing medium can increase the extrusion of sodium from cells. Presumably it is for the same reason that the retention of potassium by isolated muscles is improved when bicarbonate is present in the bathing medium.

APPENDIX 2

Physiological salines

For many purposes simple inorganic salines are required which will maintain isolated animal tissues in an active state for short periods. Some of the better known salines are listed below.

It is usually considered that to be effective in maintaining the properties of isolated tissues, a saline should have properties similar to those of the blood. The most important features are:

(1) Correct ion composition, pH, osmotic pressure and temperature.
(2) Provision of oxygen and metabolites.

Unfortunately, except in the case of the vertebrates, the blood composition of comparatively few species is known in any detail and even in the published analyses there are pitfalls for the unwary since in many cases it is not made clear whether the analyses were performed on whole blood, plasma or serum. Such information is important when devising a physiological saline as plasma proteins may bind appreciable amounts of inorganic ions, particularly divalent ions such as calcium and magnesium. Consequently, any saline made up to contain the total amount of these ions present in the blood may well contain too much of these ions in an ionised state. Many of the salines listed below have in fact been derived empirically without reference to the blood of the animal.

Notes on preparation

It is most undesirable that salines such as Tyrode which contain bicarbonate should be kept for long periods as there is danger of a loss of CO_2 and pH change. Ideally all salines should be made up freshly, shortly before use. If this is not possible, the components should be prepared separately and mixed before required. Phosphate and glucose solutions should be stored in a refrigerator to minimise bacterial action. The temperature of the saline must be raised to that of the tissue before it is added to the preparation.

Salines containing bicarbonates cannot be sterilised by boiling after they have been mixed as precipitation of calcium is then likely. Precipitation is least likely to occur during preparation of the solutions if the components are added in the order listed.

The table overleaf shows the composition of some artificial media and physiological salines for use with isolated animal tissues and organs.

All values given as grams/litre of water.

	$NaCl$	KCl	$CaCl_2$	$MgCl_2$	Na_2SO_4	$NaBr$
ARTIFICIAL SEA WATER	23·991	0·72	1·135	5·102	4·012	0·085
	23·477	0·664	1·102	4·981	3·917	—
'Ringer' solutions:						
ANNELIDS	$NaCl$	KCl	$CaCl_2$	$MgCl_2$	$MgSO_4$	$NaHCO_3$
Earthworm	6·8	0·14	0·12	—	—	0·2
Earthworm	6·0	0·12	0·2	—	—	0·1
MOLLUSCS						
Helix	5·7	0·15	1·11	—	—	—
Helix	7·0	0·3	0·1	—	0·3	1·5
Lymnaea	2·7	0·3	0·83	—	—	—
Anodonta	0·936	0·019	1·0	0·035	—	—
Mytilus	26·3	0·75	1·18	5·0	—	0·2
CRUSTACEA						
Cancer	28·7	0·91	1·63	1·16	—	0·15
Cambarus	12·0	0·4	1·5	0·25	—	0·2
INSECTS	$NaCl$	KCl	$CaCl_2$	$MgCl_2$	$NaHCO_3$	$KHCO_3$
Locusta	8·2	—	0·22	0·19	—	0·4
Periplaneta	9·32	0·77	0·5	—	0·18	—
					Na_2HPO_4	
Carausius and probably suitable for other Lepidoptera larvae	—	1·34	0·832	4·75	1·335	—
FISH						
Marine Teleosts						
Flounders	7·8	0·18	0·166	0·095	0·084	—
Lophius	12·0	0·6	0·25	0·35	0·19	—
F-W Teleosts						
						Na_2HPO_4
Salmo	7·41	0·36	0·17	—	0·31	1·6
AMPHIBIA						
Frog	6·5	0·14	0·12	—	0·2	—
				$MgSO_4$		Na_2HPO_4
Frog	4·24	0·148	—	0·146	2·1	0·356
MAMMALS	$NaCl$	KCl	$CaCl_2$	$MgCl_2$	$MgSO_4$	$NaHCO_3$
	8·0	0·2	0·2	0·05	—	1·0
	9·0	0·42	0·24	—	—	0·2

$NaHCO_3$	$SrCl_2$	H_3BO_3	NaF	KBr	pH	$Salinity$	*Author*
0·197	0·011	0·027	—	—	—	34·33	Hale (1957)
0·192	0·024	0·026	0·003	0·096	7·9–8·3	34·38	Lyman & Flemming (1940)
NaH_2PO_4							
0·01	—	—	—	—	—	—	Ambache et al. (1945)
	—	—	—	—	—	—	Prosser & Zimmerman (1943)
—	—	—	—	—	—	—	
0·5	—	—	—	—	—	—	Hedon-Fleig in Gatenby (1937)
—	—	—	—	—	—	—	Jullien & Ripplinger (1948)
—	—	—	—	—	—	—	Pilgrim (1953)
—	—	—	—	—	—	—	Hodgkin & Katz (1949)
—	—	—	—	—	—	—	Davenport (1941)
—	—	—	—	—	—	—	Van Harreveld (1936)
KH_2PO_4	NaH_2PO_4						
0·82	—	—	—	—	—	—	Hoyle (1953)
—	0·01	—	—	—	—	—	Van Asperen & Van Esch (1956)
		Glucose					
—	0·21	63·3	—	—	—	—	Wood (1957)
—	0·06	—	—	—	—	—	Forster & Hong (1958)
0·68	—	—	—	—	—	—	Young (1933)
—	0·4	—	—	—	—	—	Holmes & Stott (1960)
—	—	—	—	—	—	—	Ringer (1881)
	Na_2SO_4	*Ca gluconate*					
0·068	0·092	0·4					Boyle & Conway (1941)
NaH_2PO_4	*Glucose*						
0·04	1·0	—	—	—	—	—	Tyrode (1910)
—	1·0	—	—	—	—	—	Locke (1901)

Further reading

BAHL, K. N.: 'Excretion in the oligochaeta': *Biol. Rev. 22*, 109 (1947)

BALDWIN, E.: *An introduction to comparative biochemistry.* Cambridge University Press (1949)

BALINSKY, J. B., CRAGG, M. M. and BALDWIN, E.: 'The adaptation of amphibian waste nitrogen excretion to dehydration': *Comp. Biochem. Physiol. 3*, 236 (1961)

BARKER-JØRGENSEN, C. and LORSEN, L. O.: 'Comparative aspects of hypothalamic-hypophyseal relationships': *Ergebnisse der Biologie 22*, 1 (1960)

BEADLE, L. C.: 'Osmotic regulation and the faunas of inland waters: *Biol. Rev. 18*, 172 (1943)

BEADLE, L. C.: 'Comparative physiology: osmotic and ionic regulation in aquatic animals': *Annu. Rev. Physiol. 19*, 329 (1957)

CHEW, R. M.: 'Water metabolism of desert inhabiting vertebrates': *Biol. Rev. 36*, 1–31 (1961)

CLOUDSLEY-THOMPSON, J. L.: *Spiders, Scorpions, Centipedes and Mites.* Pergamon Press (1958)

COMAR, C. L. and BRONNER, F. (ed.): *Mineral Metabolism.* Vols. I and II, Academic Press (1960)

DENTON, E.: 'The buoyancy of marine animals': *Scientific American 203*, 118 (1960)

EDNEY, E. B.: *The water relations of terrestrial arthropods.* Cambridge University Press (1957)

FÄNGE, R., SCHMIDT-NIELSEN, K. and OSAKI, H.: 'The salt gland of the herring gull: *Biol. Bull. 115*, 162 (1958)

FARRELL, G. and TAYLOR, A. N.: 'Neuroendocrine aspects of blood volume regulation': *Annu. Rev. Physiol. 24*, 471–490 (1962)

FETCHER, E. S.: 'The water balance of marine mammals: *Quart. Rev. Biol. 14*, 451 (1939)

FLORKIN, M.: 'Metabolism et milieu chez les Crustacés': *Ann. Soc. Zool. Belge 89*, 105 (1959)

HEDGEPETH, J. W. (ed.): 'Treatise on marine ecology and palaeocology': Vol. 1, Waverley Press, Baltimore (1957)

KROGH, A.: *Osmotic regulation in aquatic animals.* Cambridge University Press (1939)

KITCHING, J. A.: 'Osmoregulation and ionic regulation in animals without kidneys, in Symposium of the Society for Experimental biology No. 8. Cambridge University Press (1954)

LEAF, A.: 'Kidney, water and electrolytes': *Annu. Rev. Physiol. 22*, 111 (1960)

LOCKWOOD, A. P. M.: ' "Ringer" solutions and some notes on the physiological basis of their chemical composition': *Comp. Biochem. Physiol. 2*, 241 (1961)

LOCKWOOD, A. P. M.: 'The osmoregulation of crustacea': *Biol. Rev. 37*, 257 (1962)

MARTIN, A. W.: 'Comparative physiology: excretion': *Annu. Rev. Physiol. 20*, 225 (1958)

NICOL, J. A. C.: *The biology of marine animals.* Pitman & Sons (1960)

PEARSE, A. S.: *The emigration of animals from the sea.* Sherwood Press (1950)

PROSSER, C. L. (ed.): *Comparative animal physiology.* W. B. Saunders Co.

RAMSAY, J. A.: *A physiological approach to the lower animals.* Cambridge University Press (1952)

RAMSAY, J. A.: 'Movements of water and electrolytes in invertebrates in Symposia of the Society for Experimental Biology 8'. Cambridge University Press

RAMSAY, J. A.: 'The comparative physiology of renal function in invertebrates in *The cell and the organism*'. Cambridge University Press (1961)

ROBERTSON, J. D.: 'The habitat of the early vertebrates': *Biol. Rev. 32*, 156 (1957)

ROBINSON, J. R.: *Reflexions on renal function.* Blackwell, Oxford (1954)

ROBINSON, J. R.: 'Metabolism of intracellular water': *Physiol. Rev. 40*, 112 (1960)

SCHMIDT-NIELSEN, K.: 'The salt secreting gland of marine birds': Circulation, *21*, 955 (1960)

SCHMIDT-NIELSEN, B., SCHMIDT-NIELSEN, K., HOUPT, T. R. and JARNUM, S. A.: 'Water balance of the camel': *Amer. J. Physiol. 185*, 185 (1956)

SHAW, J.: 'The mechanism of osmoregulation': *Comparative Biochemistry*, Vol. 2, ed. by Florkin M. and Mason, H. S. (1960)

SMITH, H. W.: 'Water regulation and its evolution in fish': *Quart. Rev. Biol. 7*, 1 (1932)

SMITH, H. W.: *The kidney: structure and function in health and disease.* Oxford (1951)

THORN, N. A.: 'Mammalian antidiuretic hormone': *Physiol. Rev. 38*, 169 (1958)

USSING, H. H.: *The alkali metal ions in biology.* Berlin (1960)

WATERMAN, T. H. (ed.): *The physiology of the crustacea*, 1: Academic Press (1960)

WILLMER, E. N.: *Tissue Culture.* Methuen (1954)

WIRZ, H.: 'Kidney, water and electrolyte metabolism': *Annu. Rev. Physiol. 23*, 577 (1961)

WRONG, O.: 'Sodium excretion and the control of extra-cellular fluid volume, in Lectures on the Scientific basis of Medicine, Vol. 8'. Athlone Press, London (1959)

Glossary

$\triangle$	The depression of freezing point of a solution below that of the pure solvent.
Euryhaline	Tolerance of wide changes in salinity.
Homoiothermic	Warm blooded.
Hyperosmotic solution	A solution whose osmotic concentration is higher than that of a reference solution.
Hyposmotic solution	A solution whose measured osmotic concentration is less than that of a reference solution.
Hypertonic solution	A solution which gains water from a hypotonic solution when separated from it by a semipermeable membrane.
Hypotonic solution	See hypertonic solution.
Isosmotic solution	A solution with the same osmotic concentration as a reference solution, e.g. isosmotic urine—urine with the same concentration as the blood.
Isotonic solution	Two solutions between which there is no passage of water when they are separated by a semipermeable membrane.
Molal	The concentration of a solution formed by dissolving the gram molecular weight of a substance in 1 kilogram of water.

Molar	The concentration of a solution formed by dissolving the gram molecular weight of a substance in water and making the volume up to one litre at 0° C.
Osmosis	The passage of water from a low to a high solute concentration through a semipermeable membrane.
Poikilothermic	Having variable body temperature.
Semipermeable membrane	A membrane permitting the passage of water but not of solute particles.
Stenohaline	Intolerant of wide changes in salinity.

Index